# Managing Gut Health

## Natural Growth Promoters as a Key to Animal Performance

Cover painting: “Hygieia” by Gustav Klimt (1862–1918).
In the Greek mythology, Hygieia is the goddess of health, symbolising enduring health and prevention of sickness.

# Managing Gut Health

*Natural Growth Promoters as a Key to Animal Performance*

Tobias Steiner

Nottingham University Press
Manor Farm, Main Street, Thrumpton
Nottingham, NG11 0AX, United Kingdom

NOTTINGHAM

First published 2006


**British Library Cataloguing in Publication Data**
Managing Gut Health
Natural Growth Promoters as a Key to Animal Performance
Steiner, T.

ISBN-10: 1-904761-45-3
ISBN-13: 978-1-904761-45-7

*Disclaimer*

Every reasonable effort has been made to ensure that the material in this book is true, correct, complete and appropriate at the time of writing. Nevertheless the publishers, the editors and the authors do not accept responsibility for any omission or error, or for any injury, damage, loss or financial consequences arising from the use of the book.

Typeset by Nottingham University Press, Nottingham
Printed and bound by Hobbs the Printers, Totton, Hampshire, England

## Table of contents

## Acknowledgement

The contribution of Dr. Christian Lückstädt to the aquaculture sections is gratefully acknowledged.

1

# INTRODUCTION

The current situation in world food supply calls for supreme efforts to ensure the increasing demand of the growing world population for staple diets and high-quality food at a cheap price. Livestock production and aquaculture are the fastest growing sectors within global agriculture. There is an increasing demand for animal products, driven by rising per capita income in fast-developing areas such as China or India. Global meat consumption, for example, is estimated to increase by 2% annually, with most of this increase occurring in developing countries. It is a major challenge to overcome the widening gap between food demand and supply, especially in developing areas. At the same time, there is growing awareness of consumers for animal welfare and food safety.

Modern intensive animal production relies on a high efficiency of feed and labour management, meaning that high numbers of animals of the same age and genetic background are kept on a relatively small unit. This, in turn, opens the door for a variety of threats and makes animals highly susceptible to stress, including immune-suppression, infectious diseases and gastrointestinal disorders such as diarrhoea, post-weaning colibacillosis or non-specific dysentery. In modern animal production operations, there is a pertinent danger created by pathogenic germs, feed contamination with mycotoxins or production of toxic gases (e.g. ammonia, hydrogen sulphide). Moreover, variation in ambient temperature and humidity, social stress and poor quality of feed and drinking water may have a detrimental impact on overall animal health. Finally, the output of nutrients such as nitrogen and phosphorus in the manure contributes to environmental pollution in areas of high-density livestock production. Thus, legislative regulations have been implemented in several countries such as Denmark, Germany or France restricting the number of livestock per hectare of land.

In recent years, the importance of gut health associated with a well-balanced gut microflora has been recognised as fundamental precondition for cost-efficient and environmentally-sound livestock production. It has been elucidated that a healthy gut is the most important precondition for transforming nutrients into performance. The major issue in modern animal nutrition is, therefore, to promote and maintain gastrointestinal health to ensure overall productivity and to deliver high-quality and safe animal products.

# 2

# GUT MICROFLORA

Microorganisms, such as bacteria, fungi or yeasts, are abundant in nature, featuring a tremendous diversity of species. Due to a steady supply of nutrients, the gut lumen represents an excellent habitat for microbial colonisation. Intensive research has been focused on the significance of a healthy digestive tract as a fundamental tool to secure animal performance. According to Conway (1994), stable gut health is based on three major factors: host physiology, diet and microflora, also referred to as the gut ecosystem (Figure 2.1). Interactions between these factors are finely balanced and any disturbance, for example caused by a drop in hygienic conditions in the production facilities or by stress, may significantly affect the gut ecosystem.

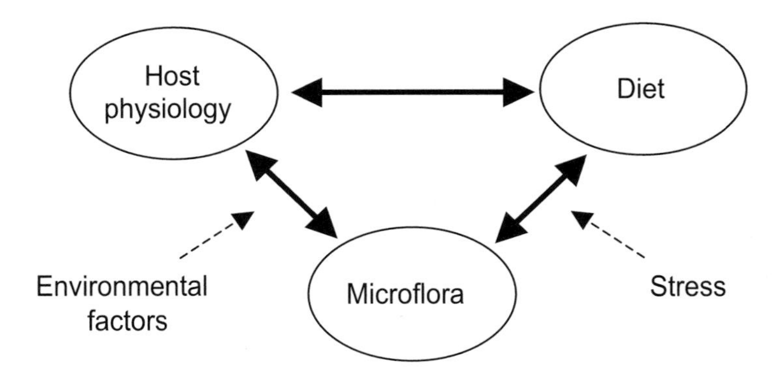


**Figure 2.1**. The gut eco system (according to Conway, 1994)

The gut microflora plays a central part within the gut ecosystem. There is a large number of microorganisms living in the gut in symbiosis with the host animal. The gut lumen and mucosal tissue are major sites of microbial colonisation. At birth, the digestive tract of neonates is devoid of microorganisms. Immediately after birth, however, as soon as the animal is exposed to the environment, a versatile microflora begins to develop. Consequently, the gut is colonised by a large number of microorganisms, most of which are harmless while others may be pathogenic. The gut microflora, both in terms of total numbers and composition, differs greatly between animal species and different gut sections. Furthermore, age and physiological state of the animal as well as diet composition may affect the gut microflora. As reviewed by Richards *et al.* (2005), the number of microorganisms present in the stomach and proximal small intestine of nonruminants (pigs, poultry) is relatively small ($10^1$-$10^5$ colony-forming units per gram of digesta, CFU/g), most of them being aerobe and facultative anaerobe bacteria such as *Lactobacilli, Streptococci* and *Bacteroides*. In poultry, the crop represents the first habitat for microbial proliferation. From the proximal to the distal part of the gastrointestinal tract, the number of microorganisms increases considerably. In the terminal ileum, population density ranges between $10^8$ and $10^9$ CFU/g of gut content. Thereafter, due to a longer residence time of the digesta, the number of microorganisms increases dramatically ($10^{10}$-$10^{12}$ CFU/g digesta) in the large intestine, the predominant species being strictly anaerobe.

The digestive system of ruminants such as cattle, sheep or goats is basically different from that of nonruminant species. The rumen is colonised by a large number of microorganisms, which allows the host to utilise indigestible feed ingredient such as plant fibre components.

The microflora may exert a number of effects in the gut, which in turn may positively or negatively impact the health status of the host animal. According to Richards *et al.* (2005), these effects can be summarised as follows:

- Production of lactic acid and fatty acids (acetate, propionate, butyrate)
- Lowering of pH
- Secretion of antimicrobial compounds (bacteriocins)
- Competition with the host and other microorganisms for nutrients
- Competition with other microorganisms for attachment sites at the gut surface
- Stimulation of host immunity
- Stimulation of cell turnover
- Increased mucus production
- Modification of intestinal morphology (villus height, crypt depth)
- Decreased fat digestibility
- Increased energy requirement of the host

In healthy animals, a stable gut microflora assists in protecting the host from pathogenic invasion. Furthermore, some bacterial species may produce antimicrobial substances, referred to as bacteriocins, which directly inhibit the growth of other microorganisms (Kelly and King, 2001; Conway, 1996). Finally, a beneficial gut microflora may stimulate the host intestinal immune response, which is yet poorly developed in young animals (Mc Cracken *et al.*, 1995; Pabst *et al.*, 1988). However, in periods of stress (e.g. gestation, farrowing, weaning, change of feeding regimen or dietary composition), there may be a shift in the microbial balance in favour of pathogenic bacteria (Ewing and Cole, 1994), bearing the risk of diarrhoea, constipation and increased gas production in the hindgut. In pigs, for example, weaning is an extremely critical point since the transition from sow's milk to solid feed is a challenge to the young animals' digestive system. Therefore, the post-weaning period is usually characterised by low feed intake and weight gain and diarrhoea coinciding with a drop in the population of *Lactobacilli* and an increase in the number of coliform bacteria (Risley, *et al.* 1992). The complex mode of action of a beneficial gut microflora has been illustrated by Ewing and Cole (1994, Figure 2.2).

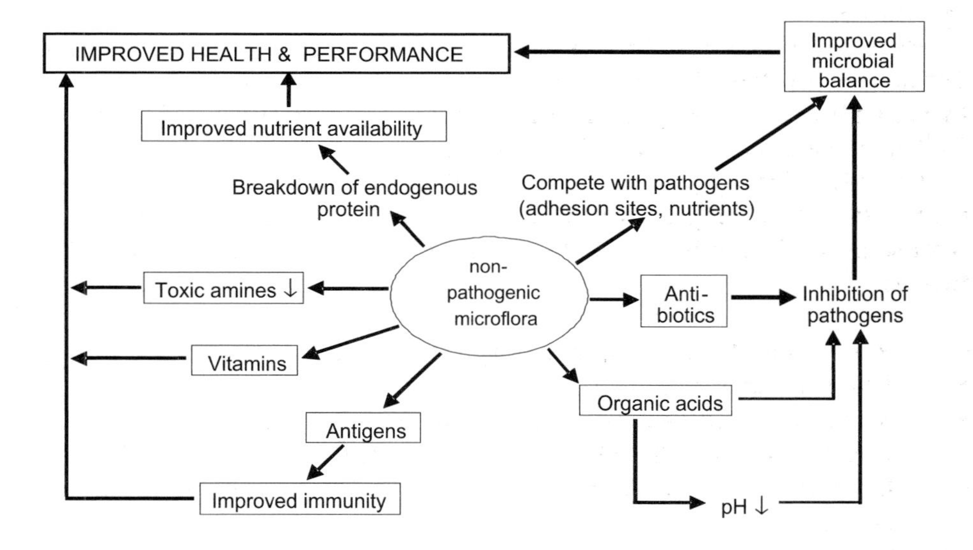


**Figure 2.2.** Effects of a non-pathogenic gut microflora (adapted from Ewing and Cole, 1994)

Healthy, adult animals are protected against pathogens by a complex immune system. The gut represents the largest organ within this defence system. However, in young piglets and newly-hatched chickens, the intestinal immune response is functionally immature (Stokes *et al.*, 2001), which makes them susceptible to pathogenic invasion during the early stage of life. There is growing evidence that the development and maturation of the gut-associated immune system is stimulated by the gut microflora. According to Bauer *et al.* (2006), oral administration of beneficial bacteria may increase the production of immunoglobulins and stimulate the activity of gut-associated immune cells (Peyer's patches).

Understanding the complex nature of the gut microflora requires highly-sensitive techniques for analysis, both in terms of quantitative and qualitative aspects. In the past, determination of microbial communities was carried out using cultivation methods. Such methods, however, are time-consuming, laborious and not sufficiently representative in many cases. At present, several molecular approaches such as polymerase chain reaction (PCR), denaturing gradient gel electrophoresis (DGGE) or fluorescent *in situ* hybridization (FISH) are available, which are considered to overcome the limitations of conventional culture techniques (Zoetendal *et al.*, 2004).

3

# USE OF ANTIBIOTIC GROWTH PROMOTERS IN ANIMAL NUTRITION

In recent decades, antibiotics have been routinely included in diets for livestock in order to prevent diseases and to increase growth performance. In Canada, for example, 90% of starter, 75% of grower and more than 50% of finisher diets for pigs contain Antimicrobial Growth Promoters (AGP) (Foote, 2003). The mode of action of AGP is still not fully elucidated. However, their growth-promoting effect can mainly be attributed to their impact on gut microflora (Collier *et al.*, 2003; Walton, 1983). Addition of AGP to diets results in a decrease in the total number of gut microorganisms (Collier *et al.*, 2003; Jensen, 1988), suppression of pathogenic bacteria and, finally, improved energy and nutrient availability for the host animal.

The positive impact of AGP on growth performance is highest in young animals, especially when hygienic conditions are poor (Taylor, 1999). AGP have been reported to reduce the incidence of several enteric disorders such as diarrhoea. However, inclusion of AGP decreased the numbers of beneficial bacteria as well (Namkung *et al.*, 2004; Engberg *et al.*, 2000). The mode of action of AGP with regard to their positive impact on growth performance can be summarised as follows:

- Suppression of (subclinical) infections
- Reduction of growth-depressing microbial metabolites (e.g. toxic amines)
- Reduced utilization of nutrients by microorganisms
- Increased nutrient uptake through a thinner intestinal wall

AGP have proven to increase growth performance in numerous trials with pigs and poultry (Gauthier, 2005). Their potential benefit in diets for pigs is shown in Figure 3.1. The greatest effects of AGP have generally been obtained in young animals which are usually more susceptible to enteric diseases than adult animals. As recommended by the Swann Committee (1969), the use of AGP only included substances which were not used in human or animal medications. Moreover, AGP have only been included in livestock feeds at sub-therapeutic levels.

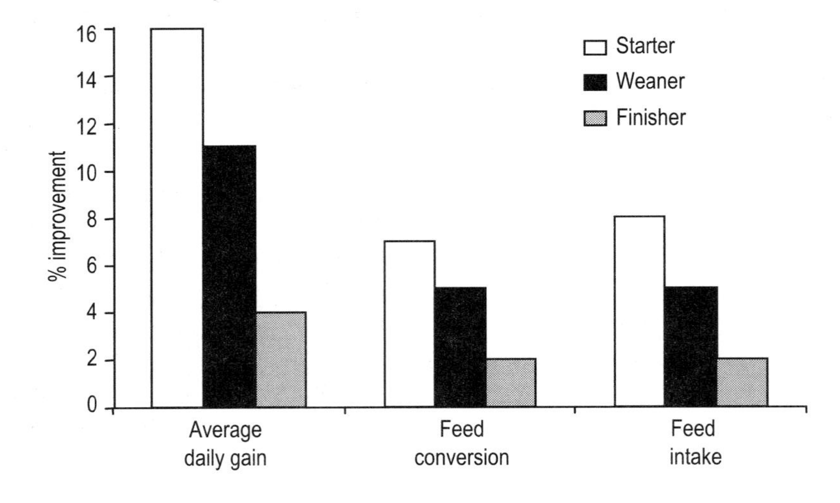


**Figure 3.1.** Impact of Antimicrobial Growth Promoters on performance parameters in different categories of pigs (data from Cromwell, 1991)

There has been growing public concern about the risk of bacterial resistance associated with the routine use of AGP in livestock production. This is indeed a serious problem, since bacteria may become resistant to antibiotics and resistance genes can be transferred rather easily among and between different bacterial species, thus bearing the risk of making antibiotic therapy ineffective in treating several human and animal diseases. The presence of resistant bacterial strains has been reported for nearly all antibiotics available on the market (Levy, 2001). Resistance genes have already been traced for *E. coli* from animals to humans. However, it has been controversially discussed whether bacterial resistances against antibiotics can be attributed to the non-therapeutic use of AGP in animal nutrition rather than to an extensive therapeutic use and misuse of antibiotics in human medicine (Domig, 2005). In 1986, Sweden was the first country that banned the use of AGP, followed by Denmark in 2000. However, the removal AGP in these countries drastically increased the use of therapeutic antibiotics, especially in piglets (Danmap, 2002). Following the examples of Sweden and Denmark, AGP were completely banned in the European Union in January 2006.

Also with respect of aquaculture, growing awareness from consumers and producers has resulted in calls for environmentally sound and sustainable aquaculture and also in the much debated shrimp production in South-East Asia (Verbeeke, 2001). Public opinion and regulation authorities in most exporting countries focus now on the misuse of antibiotics in aquaculture and public attention has been directed towards production methods (Lückstädt, 2005).

The major benefit of the removal of AGP is an expected augmentation of consumers' acceptance for products of animal origin such as meat, eggs and milk products. However, it may be speculated that the ban of AGP in livestock feeding might increase the incidence of subclinical diseases, thus resulting in reduced growth performance and, finally, increased feed costs. According to Pfirter (2003), farmers might have to face a decrease in growth performance ranging between 3 and 8% for daily weight gain, and between 2 and 5% for feed efficiency (gain/feed ratio). In any case, however, the impact of the removal of AGP on productivity will largely depend on the hygienic status established in livestock operations. Thus, the first step for the farmer is to create or maintain high hygienic standards, for example by implementing all-in/all-out systems in combination with stringent disinfection plans in pig production systems to prevent transmission of diseases between different batches of animals.

# 4

# NATURAL GROWTH PROMOTERS - ALTERNATIVES TO ANTIBIOTICS

Numerous efforts have been undertaken to develop suitable alternatives in order to counteract the anticipated drawbacks associated with the ban of AGP. Different substances often referred to as Natural Growth Promoters (NGP), have been identified as effective and safe alternatives to AGP. NGP are supposed to achieve high consumer acceptance since they do not usually pose any risk in terms of bacterial resistance or residues in animal products (Recht, 2005). At present, there is a large number of NGP available at the market, including organic acids, immune modulators, probiotics, prebiotics, feed enzymes and phytogenics. All these products have the potential to beneficially affect gut health and growth performance. Their mode of action is highly complex. However, a major common effect of all NGP is their influence on the gut microflora in qualitative or quantitative terms. The general aim of using NGP is, therefore, to establish and to maintain a well-balanced gut microflora which protects the host against pathogenic invasion.

In some cases, however, scientific reports are inconsistent regarding the efficacy of NGP, whilst their mode of action remains, at least partly, undiscovered. A lack of effects of NGP in animal experiments has sometimes been observed. Among others, level of feed intake, hygienic standards and number of animals used may be discussed as factors affecting the efficacy of NGP. The conditions under which the efficacies of NGP are determined do not necessarily reflect the situation under practical conditions. In digestibility experiments, for example, animals are often fed restrictedly, which may have an impact on passage rate or gastric emptying, when compared to *ad libitum* feeding conditions in practice. Moreover, under excellent experimental conditions, a relatively low number of animals per square metre and lower infection pressure often result in higher growth rates and feed efficiency, thus affecting the efficacy of NGP under trial conditions. Finally, the lack of consistency regarding the efficacies of NGP within one category (e.g. phytogenics, acidifiers) can be attributed to great variations in chemical composition of NGP products.

The present work provides a comprehensive review of the use of NGP as alternatives to AGP in animal nutrition. Herein, the main focus is directed to the application, potential benefits and mode of action of the following classes of NGP:

- Acidifiers
- Probiotics
- Prebiotics
- Synbiotics
- Feed enzymes
- Phytogenics
- Immune stimulants

# 5

# ACIDIFIERS

## Characterisation

At present, organic acids are considered a promising alternative to replace AGP in animal nutrition (Gauthier, 2005). Organic acids are natural constituents of plant and animal tissues and their use as feed additives is already well established. Moreover, they are produced by microbial fermentation of carbohydrates in the gut of livestock. Acidification of diets with organic acids and their salts is widely used for the purpose of feed conservation. Furthermore, growth-promoting effects of organic acids have been reported for pigs, poultry, fish and shrimp. Organic acids frequently used in animal nutrition are listed in Table 5.1.

**Table 5.1. Organic acids and salts used in animal nutrition (Mroz, 2003)**

| | *pKa value* | *Gross energy (KJ/g)* | *Corrosiveness*[1] | *Molecular weight* | *Form* |
|---|---|---|---|---|---|
| Acetic acid | 4.76 | 14.6 | ++(+) | 60.05 | Liquid |
| Benzoic acid | 4.19 | 26.4 | 0 to (+) | 122.1 | Solid |
| Butyric acid | 4.82 | 24.8 | + | 88.12 | Oily liquid |
| Citric acid | 3.14/4.80/6.40 | 10.2 | 0 to ++ | 192.1 | Solid |
| Formic acid | 3.75 | 5.7 | +++ | 46.03 | Liquid |
| Fumaric acid | 3.02/4.38 | 11.5 | 0 to (+) | 116.1 | Solid |
| Lactic acid | 3.86 | 15.1 | (+) | 90.08 | Liquid |
| Malic acid | 3.46/5.10 | 10.0 | (+) | 134.1 | Solid |
| Propionic acid | 4.88 | 20.6 | ++ | 74.08 | Oily liquid |
| Calcium formate | | 11.0 | 0 | 130.1 | Solid |
| Calcium butyrate | | 48.0 | 0 | 214.0 | Solid |
| Calcium lactate | | 30.0 | 0 | 308.3 | Solid |
| Calcium propionate | | 40.0 | 0 | 184.1 | Solid |
| Potassium diformate | | 11.4 | 0 | 130.0 | Solid |
| Magnesium citrate | | 10.0 | 0 | 214.4 | Solid |
| Sodium benzoate | | 26.0 | 0 | 144.1 | Solid |
| Sodium lactate | | 15.0 | 0 | 112.1 | Solid |

[1]Corrosiveness: 0 negligible, + low, ++ medium, +++ high

Due to their corrosive nature, care must be taken in handling organic acids. Solid acidifiers are generally less corrosive than liquid acids.

Few studies evaluated the efficacy of inorganic acids such as phosphoric, sulphuric or hydrochloric acid in nonruminant nutrition (Mahan *et al.*, 1999; Schoenherr, 1994; Giesting, 1986). Phosphoric acid is frequently used in combination with organic acids, since other inorganic acids have caused a marked decrease in growth performance (Giesting, 1986). However, positive effects on performance parameters have also been reported with hydrochloric acid in weaning piglets (Mahan *et al.*, 1999). Phosphoric acid assists in lowering the pH values. Furthermore, it is usually cheaper as compared to organic acids and, in addition, represents a source of highly-available phosphorus.

Pure acidifiers are usually rapidly utilised in the upper part of the digestive tract. Attachment of acidifiers to inorganic (e.g. silicates, vermiculite) or organic carriers (e.g. fructooligosaccharides), however, results in sequential release of the acids along the gastrointestinal tract, thus promoting a long-lasting efficacy of acidifiers in the stomach and gut.

## Mode of action

The mode of action of organic acids and their salts is rather complex and, at least partly, though not fully, elucidated. It is generally accepted that organic acids and their salts suppress the growth of pathogenic microorganisms, both in the feed and in the gastrointestinal tract of livestock. As outlined by Roth and Kirchgessner (1995), acidifiers exert their beneficial effects at three different levels, i.e. feed, gut and intermediary metabolism (Table 5.2). The main effects are discussed in the following paragraphs.

**Table 5.2. Mode of action of organic acids and salts (adapted from Roth and Kirchgessner, 1995)**

| | | |
|---|---|---|
| Raw materials/finished feed | | Lowering of pH<br>Reduction in buffering capacity<br>Antimicrobial effect |
| Gastrointestinal tract | 1) Proton ($H^+$) | Lowering of pH (mainly in stomach)<br>Improved efficacy of pepsin<br>Antimicrobial effect |
| | 2) Anion | Antimicrobial effect<br>Complexing agent for cations ($Ca^{2+}$, $Mg^{2+}$, $Fe^{2+}$, $Cu^{2+}$, $Zn^{2+}$) |
| Intermediary metabolism | | Utilisation as energy source |

## EFFECTS OF ACIDIFIERS ON MICROORGANISMS IN FEED AND GASTROINTESTINAL TRACT

As shown in Table 5.2, the suppressive impact of organic acids on different microorganisms is based on several mechanisms. One main effect of organic acids is a reduction of the pH values in the feed as well as in the gastrointestinal tract of livestock, thus creating unfavourable conditions for potentially pathogenic microorganisms. It has been recognised that the growth rates of pathogenic bacteria such as *E. coli* or *Salmonella* may be decreased at low pH values, while beneficial bacteria (e.g. *Lactobacilli*) are usually not inhibited in their growth (Kirchgessner and Roth, 1988).

Even under good hygienic conditions, raw materials and finished feeds contain a certain number of moulds, bacteria and yeasts. Some mould species (e.g. *Aspergillus*, *Penicillium*, *Fusarium*) are known to produce mycotoxins, which may cause severe health problems, including, among many others, immune-suppression, fertility problems, digestive disorders and splay legs. Conservation of feedstuffs and diets with organic acids has proven to reduce the microbial counts in the feed, thus avoiding degradation and maintaining feed safety (Eidelsburger, 1997). As shown by Adams (2001), preservation with a mixture of propionic and formic acid at 3 kg/t successfully reduced the initial contamination of wheat with *Salmonella enteritidis* by 59%.

A decrease in gastrointestinal pH may serve as a barrier against pathogenic microorganisms ingested via feed, faeces or straw. According to Namkung *et al.* (2004) and Walsh *et al.* (2003), acidification of diets resulted in a reduction in the number of coliforms or *E. coli*, respectively, in the faeces of piglets. This is in agreement with findings by Biagi *et al.* (2003) who indicated that addition of fumaric acid to diets for piglets decreased the population of clostridia in the small intestine as well as counts of *Clostridia* and coliforms in the caecum.

As reported by Radcliffe *et al.* (1998), addition of 1.5 or 3.0% citric acid decreased the original pH value in the feed from 6.5 to 5.0 and 4.4, respectively. In a study by Rice *et al.* (2002), addition of citric acid (3%) to corn-soybean meal diets decreased the gastric pH in pigs by 11%. According to Eckel *et al.* (1992), supplementation of diets with 0.6, 1.2 or 1.8% formic acid successfully reduced the incidence of diarrhoea in pigs. Consequently, diet acidification is effective in preventing enteric diseases and digestive disorders, primarily during the post-weaning period in piglets. In experiments with piglets, Kirchgessner *et al.* (1992) showed that supplementation of a control diet with calcium formiate or formic acid reduced the numbers of *E. coli*, *Enterococci* and *Bacteroides* species in the duodenum. Similar effects were reported by Canibe *et al.* (2001), Overland *et al.* (2000) and Bolduan *et al.* (1988). In contrast, several studies did not reveal significant effects of diet acidification on pH in different sections of the gastrointestinal tract (e.g. Namkung *et al.*, 2004; Gabert and Sauer, 1995; Roth *et al.*, 1992a), presumably because digesta pH may have been confounded due to the application of slaughter methods rather than cannulation techniques.

Apart from their effect on pH conditions, there is also a direct antimicrobial effect of organic acids. In the non-dissociated form, the acid molecules are able to penetrate through the semi-permeable bacterial cell wall. Inside the cell cytoplasm, the acids dissociate, thereby lowering the intracellular bacterial pH (Figure 5.1). Consequently, the cell starts to utilise energy to restore its original pH. Finally, the bacterial metabolism is disrupted by inhibition of cell enzymes and nutrient transport systems (Lueck, 1980).

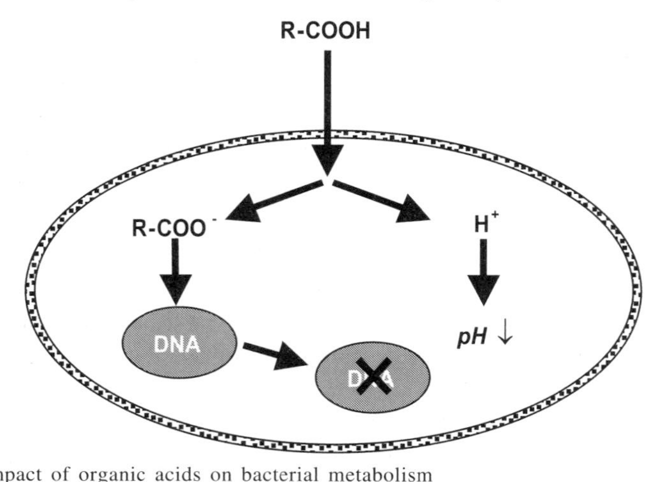


**Figure 5.1.** Impact of organic acids on bacterial metabolism

The efficiency of organic acids in suppressing bacterial growth depends on their $pK_a$ value, which is the pH value at which 50% of the acid is dissociated. As shown in Table 5.1, $pK_a$ values of various organic acids range between 3.14 (citric acid) and 4.88 (propionic acid). Low $pK_a$ values generally indicate a strong pH-decreasing impact of a respective acid, whereas a distinct antimicrobial effect is obtained at higher $pK_a$ values. The combination of different acidifiers with different $pK_a$ values may, therefore, exert synergistic effects and maximise the beneficial impact of supplementation with organic acids on gut health and growth performance (Piva *et al.*, 2002).

Propionic acid strongly inhibits the growth of moulds. It is, therefore, mainly used as preservative during feed storage (Foegeding and Busta, 1991). Formic acid mainly inhibits the growth of yeasts and bacteria such as *E. coli* or *Salmonella*, whereas *Lactobacilli* and moulds are relatively resistant against formic acid (Lueck, 1980). The inhibition potency of organic acids for *Salmonella typhimurium* was reported to increase in the following order: acetic < formic < propionic < lactic < sorbic < benzoic acid (Jensen *et al.*, 2001). For coliform bacteria, Mroz (2003) reported a different order: propionic < formic < butyric < lactic < fumaric < benzoic acid. As indicated by Maribo *et al.* (2000), administration of lactic acid may increase the population of *Lactobacilli* in the gut of pigs while decreasing the number of coliform species.

## IMPACT OF ACIDIFIERS ON PROTEIN DIGESTION

Especially in young piglets after weaning, protein digestion is limited because the secretion of hydrochloric acid in the stomach, which is necessary for the activation of pepsinogen to initiate protein digestion, is yet insufficient (Cranwell, 1985). However, young animals require high dietary concentrations of protein and minerals, resulting in a high buffering capacity of the feed along with a reduced efficiency of protein and amino acid digestibility (Blank *et al.*, 1999). Poor gastric acidification of the stomach digesta, therefore, is a cause of bacterial overgrowth in the small and large intestine, thus increasing the incidence of diarrhoea in young piglets.

Acidification of diets with organic acids improves gastric proteolysis by reducing the pH and, therefore, results in increased efficacy of pepsin. As reported by Radcliffe *et al.* (1998), supplementation of diets with citric acid reduced the gastric pH from 3.8 to 3.3 in pigs. Consequently, microbial dysfermentation of undigested protein in the large intestine is reduced, coinciding with a lower production of ammonia and toxic amines such as putrescine or cadaverine (Blank *et al.*, 1999). All these effects contribute to a higher protein digestibility and improved feed efficiency. As reported by Mroz *et al.* (2002), dietary acidification with formic acid or formiates may also reduce the amounts of faeces and urine voided by growing-finishing pigs.

Moreover, it has been hypothesised that acidification of diets may reduce the rate of gastric emptying. Presumably, a lower rate of gastric emptying might increase the time that is available for protein hydrolysis in the stomach, thus resulting in improved protein and amino acid digestibility. At present, however, scientific proof for this hypothesis is still missing, since most studies failed to show significant effects of acidifiers on gastric emptying (e.g. Gabert and Sauer, 1995; Risley *et al.*, 1992).

## ADDITIONAL EFFECTS OF ACIDIFIERS

As indicated by Kirchgessner and Roth (1982), several cations such as calcium, phosphorus and zinc may be bound by the acid anion, thus causing an increase in apparent digestibility of these elements. In studies with pigs, coefficients of apparent total tract digestibilities of phosphorus and calcium were significantly increased by up to 5 and 6%, respectively, when a control diet based on barley, soybean meal, corn and tapioca was supplemented with lactic acid (Jongbloed *et al.*, 1996b). Supplementation of the control diet with formic acid significantly improved apparent total tract digestibilities of phosphorus and calcium by 5 and 8%, respectively.

Furthermore, organic acids and their salts may serve as energy sources that are utilised in the intermediary metabolism, thus contributing to the energy balance of the animal. As shown in Table 5.1, the energy content varies considerably between different organic acids (5.7-48.0).

Finally, it is speculated that organic acids may directly modulate the development of mucosal cells and, therefore, contribute to a rapid recovery of the gut epithelium and probably an increase in the overall absorptive surface (Blank *et al.*, 1999). Moreover, Thaela *et al.* (1998) reported that pancreatic secretions were stimulated by lactic acid supplementation in piglets. Thus, acidification of diets for starter or weaner pigs may be a suitable tool for improving the secretion of endogenous pancreatic enzymes such as trypsin or chymotrypsin in young animals.

## Effects of acidifiers on performance

Several reports have confirmed that organic acids effectively increase growth performance in different animal species. As reviewed by Freitag *et al.* (1999), supplementation of diets for pigs and poultry may improve weight gain and feed efficiency by 3-15 and 2-10%, respectively. The response of performance parameters mainly depends on the dosage of a respective acid as well as on the inherent buffering capacity of the feed and the composition of acidifier products. Furthermore, the effects of acidifiers are supposed to be higher in young animals as compared to adult animals.

### EFFECTS OF ACIDIFIERS IN PIGS

Most of the research on organic acid feed supplements was focused on their application in pigs, where supplementation of diets with organic acids and their salts may improve energy and protein digestibility by up to 2 and 4%, respectively. According to Partanen (2001), however, ileal digestibility of lysine may be improved by up to 8 and 5% in piglets and growing pigs, respectively. The highest responses were obtained in piglets during the first three weeks of age, probably owing to the deficiency in gastric secretion of hydrochloric acid.

Different organic acids were tested for their potential to increase growth performance and feed efficiency in different categories of pigs. As reported by Giesting *et al.* (1991), supplementation of soybean meal-based control diets for starter pigs with fumaric acid at 2 or 3% of finished feed improved average daily gain and feed conversion rate by up to 29 ($P > 0.05$) and 30% ($P < 0.05$), respectively. In studies with piglets, supplementation of corn-soybean meal diets with citric acid linearly increased average daily gain and feed efficiency (Radcliffe *et al.*, 1998). Moreover, addition of citric acid at 3 and 6% of finished feed significantly improved feed efficiency in 11-kg piglets by 10 and 9%, respectively (Boling *et al.*, 2000). However, average daily gain was not affected in this study. Positive effects on growth performance were also obtained by Jongbloed *et al.* (1996b), Krause *et al.* (1994) and Kornegay *et al.* (1994).

## EFFECTS OF ACIDIFIERS IN POULTRY

Only few experiments have been carried out so far that determined the impact of acidifiers on growth performance of poultry. The efficacy of a commercial acidifier containing a blend of propionic acid and formic acid based on an inorganic carrier has been investigated in a trial with 240 birds. In this trial, broilers were fed a basal diet either supplemented or not supplemented with the acidifier at 3 g/kg (Lückstädt *et al.*, 2004). As shown in Table 5.3, the acidifier successfully improved performance parameters, as investigated after three experimental weeks. After five weeks, average daily body weight of the birds fed the acidifier tended ($P = 0.06$) to be higher as compared to birds fed the control diet (1759 vs. 1662 g).

**Table 5.3. Effect of Biotronic® (blend of propionic and formic acid) in broilers after three experimental weeks (Lückstädt *et al.*, 2004)**

| | *Dietary treatment* | | *P value* |
|---|---|---|---|
| | *Control* | *+ Acidifier* | |
| Average body weight (g) | 731 | 773 | 0.01 |
| Average daily weight gain (g) | 52 | 57 | 0.01 |
| Average daily feed intake (g) | 76 | 81 | 0.05 |
| Feed conversion rate | 1.30 | 1.29 | n.s. |

Finally, the use of the acidifier resulted in an overall economic benefit, resulting from the improvement in bird performance. Skinner *et al.* (1991) obtained increased live weight (after 49 days) and feed intake in studies with broilers, when diets were supplemented with fumaric acid at dosages ranging between 1.25 and 5.0 g/kg. Improved growth performance in poultry was also reported by Vogt *et al.* (1981). Finally, Patten and Waldroup (1988) reported that addition of 0.5 or 1.0% of fumaric acid to diets for broilers significantly increased body weight.

## EFFECTS OF ACIDIFIERS IN AQUACULTURE

Since the use of fish silage from preserved fish and fish viscera included preservation with acids (Åsgård and Austreng, 1981; Gildbert and Raa, 1977), this group of additives came under scientific observation in fish as well, first with carnivore species like Rainbow trout (*Oncorhynchus mykiss*), Atlantic salmon (*Salmo salar*) and Artic charr (*Salvelinus alpinus*), but also with herbivorous filter feeders such as tilapia, and also shrimp.

Ringø (1991) fed Arctic charr commercial diets with or without supplemental sodium salts of lactic and propionic acid (1%). Fish fed the diet with added sodium lactate increased their weight from around 310 g to about 630 g within 84 days of the experiment, while the difference from the negative control group (final weight of fish: 520 g) was significant. The inclusion of sodium propionate, however, had a growth-depressing effect compared to the control. The gut content from Arctic charr fed the sodium lactate-supplemented diet contained significantly lower amounts of water, energy, lipids, protein and free amino acids. It was observed that in intensive feeding regimens, as often appear under aquaculture conditions, charr may tend to increase the incidence of diarrhoea. When charr were fed sodium lactate, no nutritive diarrhoea appeared, probably because of much lower amounts of remaining nutrients and water in the gut. Furthermore, it was assumed that the growth-promoting effect of dietary lactate in Arctic charr is caused by the relatively slow gastric emptying rate (Gislason *et al.*, 1996). An increased retention time in the stomach augments the antibacterial potential of the lactic acid salt and can, therefore, inhibit colonisation of pathogenic bacteria (Sissons, 1989).

Feeding sodium lactate to Atlantic salmon juveniles (15 kg/t), however, did not show such a prolonged effect (Gislason *et al.*, 1996; Ringø *et al.*, 1994) compared to charr. Ringø *et al.* (1994) found slightly increased survival rates in salmon fed lactate (85% compared to 80%), while Gislason *et al.* (1996) determined a higher growth rate. However, none of those differences were statistically significant. These findings may suggest that the influence of lactate is a result of differences in the digestive physiology between the two fish species, for instance a longer retention time of lactate in the stomach in charr. However, lower bacterial challenge, due to the use of the organic acid salt, may have tended to increase survival rates.

Recently, a trial with organic acid salts was carried out with Rainbow trout (de Wet, 2005, Table 5.4). This study aimed to evaluate an organic acid blend (5-15 kg/t), mainly consisting of formate and sorbate, for its use in trout nutrition to improve performance parameters in comparison to an AGP (40 ppm Flavomycin). Rainbow trout fingerlings were kept in flow-through ponds and fed three times daily to apparent satiety. As shown in Table 5.4, fish feeding on 10 and 15 kg/t had significantly higher final weights compared to the negative control group, while there was no difference in the group treated with AGP. Feed conversion ratio tended to be lower with increasing dosages of the acid blend, even if compared to the AGP group. It can be derived from the results of this study that the application of the acidifier had the potential to improve weight gain and feed conversion ratio in trout by 20 and 15%, respectively. This data proved that inclusion of organic acids is suitable to replace AGP in Rainbow trout grower feeds.

**Table 5.4. Influence of an organic acid blend in comparison to an antibiotic growth promoter (AGP) on growth performance in Rainbow trout (adapted from de Wet, 2005)**

| *Parameter* | *Control* | *AGP* | *Acidifier (kg/t)* | | |
|---|---|---|---|---|---|
| | | | *5* | *10* | *15* |
| Initial weight (g) | 40.3 | 42.3 | 40.0 | 37.3 | 37.2 |
| Final weight (g) | $184.8^{a}$ | $235.4^{b}$ | $205.6^{ab}$ | $231.2^{b}$ | $231.4^{b}$ |
| FCR | 1.22 | 1.10 | 1.09 | 1.08 | 1.04 |
| Specific growth rate (%) | $1.23^{a}$ | $1.37^{b}$ | $1.23^{a}$ | $1.29^{ab}$ | $1.37^{b}$ |
| Survival (%) | 82.7 | 88.8 | 85.0 | 85.8 | 89.6 |

$^{ab}$within rows, means without common superscripts are significantly different ($P < 0.05$).

The use of organic acids, however, was not only tested in Salmoniformes, but also in tropical warm-water species such as tilapia. Ramli *et al.* (2005) evaluated potassium-diformate at different concentrations (0, 2, 3 and 5 kg/t) as an NGP in tilapia grow-out. Fish were challenged orally starting on day 10 of the culture period with *Vibrio anguillarum* at 105 CFU per day over a period of 20 days. Over the whole feeding period from day 1 to day 85, potassium-diformate significantly increased weight gain and feed efficiency. Survival rates after the challenge with *V. anguillarum* were also significantly higher in fish fed the acidified diets as compared to the negative control.

Furthermore, the effect was dose-dependent. The 2 kg/t inclusion of the potassium salt of the formic acid lead to an improvement in weight gain and feed conversion ratio by 19 and 8%, respectively, indicating that the acidifier is able to counteract bacterial infections in tilapia.

Based on the above-mentioned studies and trials, it can be concluded that the use of organic acid salts or acid blends is a promising option to promote the performance of a wide variety of aquaculture species. It is furthermore suggested that the impact of bacterial infections can be reduced, thereby leading to higher survival rates. The use of acidifiers in aquaculture can, therefore, be an efficient tool to achieve sustainable and economical fish and shrimp production.

## EFFECTS OF ACIDIFIERS IN RUMINANT NUTRITION

Conservation of forages such as grass, corn or cereals is commonly carried out in dairy and beef cattle operations. The principal aim of ensiling feed is to prevent loss of nutrients by creating favourable conditions for the growth of acetic and lactic acid-producing bacteria at the expense of undesired microorganisms such as yeasts, moulds, clostridia or species of the *Coli aerogenes* group. Appropriate conservation is generally achieved by a rapid decline of the pH in the feed bulk resulting from fermentation of carbohydrates predominantly by lactic acid bacteria. However, the overall success of ensiling depends on different factors such as initial moisture, concentration of readily fermentable carbohydrates and protein, composition of the epiphytic microflora as well as degree of contamination. Since silages are easily spoiled by aerobic yeasts and moulds, rapid removal of the residual oxygen through quick filling and immediate sealing of the silos is mandatory. The faster the fermentation is completed, the more nutrients are finally retained.

Thus, in ruminant nutrition, organic acids can be used as silage additives in order to improve the stability of silages during storage. Silage additives based on different organic acids have been shown to improve the aerobic stability of the conserved feed (Kung *et al.*, 1998; Sebastian *et al.*, 1996; Huber and Soejono, 1977). These additives are added at ensiling to prevent microbial nutrient degradation and spoilage, thus contributing to the overall quality of the feed, especially under poor hygienic conditions and high moisture contents. Propionic acid is usually the predominant active ingredient in such preparations (Ranjit and Kung, 2000). However, since organic acids may be rather expensive, the use of microbial silage inoculants (Chapter 6) as alternative to acidifiers may be more feasible under economic points of view.

6

# PROBIOTICS

## Characterisation

Since the importance of a well-balanced gut microflora for adequate health and high performance has been recognised, feeding strategies have been directed to control the microbial gastrointestinal environment by nutritional means. One key strategy is to feed directly the microorganisms which are supposed to exert beneficial effects in the gut. Probiotics are live microorganisms which are supplemented to the feed in order to establish a beneficial gut microflora (Fuller, 1989). Thus, probiotics have the potential to beneficially affect gut health by modification of the gut microflora, especially in young animals, in which a stable gut microflora is not yet established. As will be discussed later, positive effects on growth performance have been reported in pigs, poultry, fish, shrimps and ruminants. The use of probiotics has a long tradition in human nutrition, where lactic acid-producing microorganisms such as *Lactobacilli* or *Bifidobacteria* are frequently included in milk products (e.g. yoghurts, kefir). In the United States, probiotic feed additives are referred to as "direct-fed microbials".

## Required properties of probiotics

Since the beneficial effects of probiotics are based on a modification of the gut microflora, it is a prerequisite that the microorganisms reach the gut in a viable form. Thus, their survival during feed processing, storage and passage through the highly acidic stomach is a fundamental requirement which is in some cases difficult to achieve. Probiotics may differ substantially in their ability to withstand high temperatures as well as low pH values. *Bacillus* spores, for example, are highly heat-resistant. In contrast, other bacteria (e.g. *Enterococcus faecium*) are rapidly inactivated during pelleting, where temperatures can exceed 80 °C (Figure 6.1). At present, the stability of probiotics may be improved by different technological treatments such as coating or absorption into globuli (Simon, 2005).

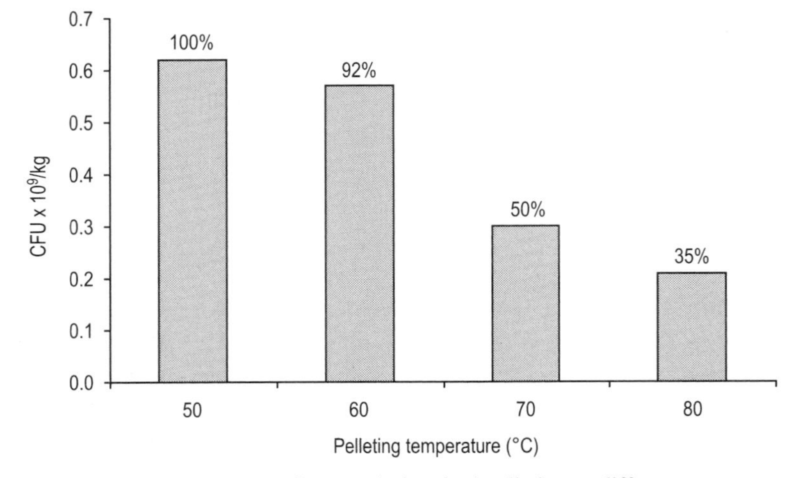


**Figure 6.1.** Stability of *Enterococcus faecium* during feed pelleting at different temperatures (Simon, 2005)

## Microorganisms used as probiotics in animal nutrition

In human nutrition, *Lactobacilli* and *Bifidobacteria* are frequently included in yoghurts and other milk products. However, due to their poor stability during storage, their application in animal nutrition is rather limited. Probiotic feed additives generally consist of one single strain or a combination of several strains of bacteria, *Bacillus* spores or yeast species (multi-strain). Preparations authorised for use in animal nutrition in the European Union include different strains of *Enterococcus*, *Bacillus*, *Lactobacillus*, *Pediococcus* or *Saccharomyces* (Table 6.1).

**Table 6.1. Bacterial and yeast strains used as probiotics in the European Union (adapted from Simon, 2005 and Lee *et al.*, 1999)**

| | |
|---|---|
| Bacteria | Bacteria |
| *Bacillus cereus* | *L. farciminis* |
| *B. licheniformis* | *L. casei* |
| *B. subtilis* | *L. plantarum* |
| *Bifidobacterium bifidum* | *L. rhamnosus* |
| *B. breve* | *Pediococcus acidilactici* |
| *B. pseudolongum* | *Streptococcus infantarius* |
| *B. thermophilum* | |
| *Enterococcus faecium* | Yeasts |
| *Kluyveromyces marxianus* | *Saccharomyces cerevisiae* |
| *Lactobacillus acidophilus* | *S. boulardi* |

## Mode of action

As described in more detail in the following paragraphs, the mode of action of probiotic feed additives is mainly based on three principles:

- Competitive exclusion
- Bacterial antagonism
- Immune modulation

The benefit of probiotics with respect to health status and performance is expected to be highest in young animals such as piglets, newly-hatched chickens or calves, because these animals have not yet developed a stable gut microflora. Moreover, when animals undergo therapeutic treatment of diseases with antibiotics, the gut microflora is generally decimated. Therefore, administration of probiotics after antibiotic treatment assists in re-establishing a beneficial gut microflora to prevent the host from recurrent pathogenic colonisation.

### COMPETITIVE EXCLUSION

The concept of competitive exclusion indicates that cultures of selected, beneficial microorganisms (Table 6.1), supplemented to the feed, compete with potentially harmful bacteria in terms of adhesion sites and organic substrates (mainly carbon and energy sources). Probiotics may colonise and multiply in the gut, thereby blocking receptor sites and preventing the attachment of other bacteria including harmful species such as enteropathogenic *E. coli* or *Salmonella*. Moreover, microorganisms present in the gut require nutrients originating from ingested feed. Thus, it can be hypothesised that a beneficial microflora will inhibit pathogens by competing for available nutrients. However, since the rate of gastrointestinal passage is relatively high, it remains rather speculative if nutrient availability is actually limiting the growth of microorganisms in the gut. Undoubtedly, probiotics have the potential to decrease the risk of infections and intestinal disorders. As shown in *in vitro* studies by Hillman *et al.* (1995), growth of enterotoxic *E. coli* was successfully inhibited by different strains of *Lactobacilli*. Pascual *et al.* (1999) indicated that *Lactobacilli* had a great capacity to adhere to the gut epithelium of broilers. In a study with one-day-old Leghorn chickens, oral administration of *Lactobacillus salivarius* successfully inhibited colonisation of *Salmonella enteritidis* in the gut of the experimental birds (Pascual *et al.*, 1999). As reported by Berchieri *et al.* (2006), a combination of different lactic acid bacteria significantly reduced the levels of *Salmonella enteritidis* in caecal contents of broilers which had been orally inoculated with the pathogen. In piglets, attachment of enterotoxic *E. coli* to the small intestinal epithelium was inhibited by dietary supplementation with *Enterococcus faecium* (Jin *et al.*, 2000).

Different studies with piglets and growing pigs (Shu *et al.*, 2001; Kyriakis *et al.*, 1999; Underdahl, 1983) showed that addition of probiotics to feed significantly decreased the incidence and severity of diarrhoea. Moreover, administration of probiotics to sows during gestation and lactation may also affect growth performance of piglets as well as the body condition of sows during and after lactation. In a study by Alexopoulos *et al.* (2004), sows were fed diets supplemented with or without *Bacillus* spores, beginning 14 days before farrowing and ending at weaning. Addition of the probiotic reduced the incidence of diarrhoea and mortality rate, and increased the body weight of the piglets. Moreover, probiotic supplementation stimulated the sows' feed intake *post partum* and reduced their body weight loss during lactation. Similarly, Abu-Tarboush *et al.* (1996) and Maeng *et al.* (1987) reported that probiotic supplementation (*Lactobacillus acidophilus* or *Streptococcus faecium*) significantly reduced the incidence of scouring in calves.

## BACTERIAL ANTAGONISM

Apart from competitive exclusion, there is a second mode of action by which probiotics may prevent the host from pathogenic overgrowth. Probiotic microorganisms, once established in the gut, may produce substances with bactericidal or bacteriostatic properties (bacteriocins) such as lactoferrin, lysozyme, hydrogen peroxide as well as several organic acids. These substances have a detrimental impact on harmful bacteria, which is primarily due to a lowering of the gut pH (Kelly and King, 2001; Conway, 1996). A decrease in pH may partially offset the low secretion of hydrochloric acid in the stomach of weanling piglets. In addition, competition for energy and nutrients between probiotic and other bacteria may result in a suppression of pathogenic species (Ewing and Cole, 1994). Finally, the stimulation of enzymes, activation of macrophages and anti-tumour activity associated with probiotics may contribute to the health status and promote growth performance of the host animal (Walsh *et al.*, 2004). The beneficial impact of probiotics in comparison to antibiotic supplementation on caecal microflora of broilers has been demonstrated by Mountzouris *et al.* (2006). The results of this trial are presented in Figure 6.2. In total 400 one-day old broilers were fed corn-soybean meal-based control diets with or without supplementation of either a multi-strain probiotic feed additive based on *Lactobacilli*, *Bifidobacteria*, *Enterococcus* and *Pediococcus* or a commercial AGP (Avilamycin). Compared to the control and AGP treatment, the probiotic additive significantly increased the numbers of *Bifidobacteria*, *Lactobacilli* and Gram-positive cocci. Moreover, growth performance in birds fed supplemental probiotics was similar ($P > 0.05$) as compared to birds fed the AGP.

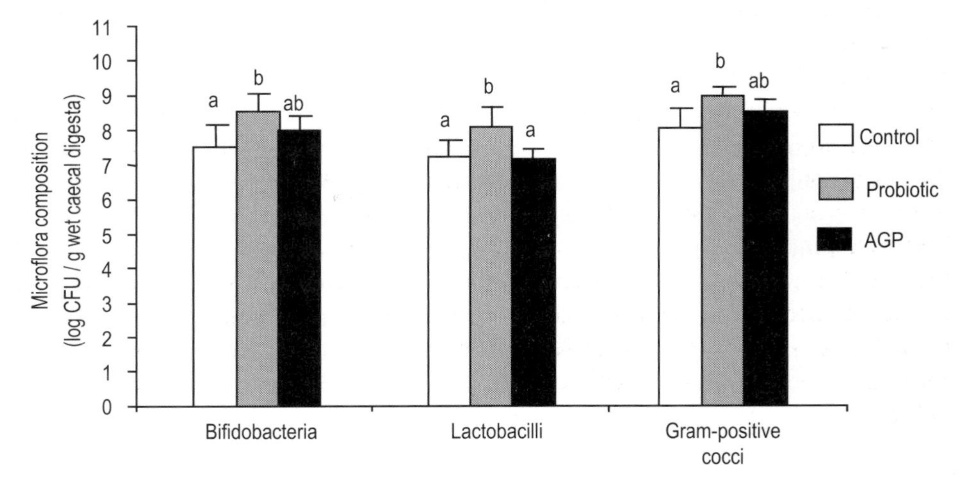


**Figure 6.2.** Impact of probiotics (Biomin® PoultryStar) and antimicrobial growth promoters (AGP) on caecal counts of different bacterial species (adapted from Mountzouris *et al.*, 2006)

Additionally, specific microorganisms are known to eliminate the negative effects of undesired feed ingredients such as mycotoxins. The anaerobic bacterium *Eubacterium BBSH 797*, for example, has been shown to degrade trichothecenes (Fuchs *et al.*, 2002), whereas a newly-discovered yeast strain, *Trichosporon mycotoxinivorans* is capable of detoxifying ochratoxins and zearalenone (Molnar *et al.*, 2004).

In conclusion, probiotics have the potential to promote and stabilise a well-balanced, beneficial gut microflora, which makes them a promising alternative to AGP, particularly in young poultry, piglets and sows.

## IMMUNE MODULATION

The gut represents the largest immune organ in mammals and there is a distinct interaction between the gut microflora and the immune system. The development and activation of the humoral and cellular gut-associated immune system is largely affected by the development of the gut microflora

(Cebra, 1999). According to Lan *et al.* (2005), microbial communities can support the animal's defence against invading pathogens by stimulating gastrointestinal immune response. The stimulatory impact of lactic acid-bacteria on the immune status has been reviewed by Perdigón *et al.* (2001). As will be discussed in more detail in Chapter 10, cell walls from yeasts or specific bacteria may promote the activity of macrophages or induce systemic immune responses.

## Effects of probiotics on performance

Effects of supplementation with probiotics on performance parameters in pigs were summarised in a review by Simon (2005), including results of 22 publications. It can be derived from this review that, in most cases, supplementation with probiotics tended to improve average daily gain and feed efficiency. However, improvements in these parameters were significant only in 7 or 5%, respectively, of the experiments that were reviewed.

In general, the lack of significant effects of probiotic supplementation on performance parameters may be attributed to a number of variable factors, including age of animals, dose of probiotics, type of diet, microbial genus and species as well as viability of probiotic strains during storage, feed processing and gastric passage. Furthermore, high variation within treatment groups depending on the microbial status of individual animals may have masked possible effects of probiotic supplementation. Furthermore, most studies are carried out under excellent hygienic conditions, which may not necessarily reflect the situation under practical conditions. Thus, it can be assumed that the gut microflora of the animals is already well balanced in many experiments, probably leaving insufficient room for further improvement through probiotics.

### EFFECTS OF PROBIOTICS IN PIGS

Probiotics were tested in different categories of pigs. According to Zani *et al.* (1998), supplementation of diets with a probiotic preparation based on *Bacillus cereus* significantly improved daily weight gain and feed conversion ratio by 24 and 19%, respectively. In another trial with piglets (Kyriakis *et al.*, 1999), addition of probiotics originating from *Bacillus licheniformis* to a control diet improved average daily weight gain as well. Moreover, the incidence and severity of diarrhoea as well as mortality rate were significantly decreased in a study by Kyriakis *et al.* (1999). Shim (2005) fed a two- or multi-strain probiotic feed additive to growing pigs and observed significant improvements in average daily weight gain and feed efficiency as well. Additionally, administration of the multi-strain probiotic tended to cause higher weight gain and feed efficiency in comparison to the two-strain probiotic, indicating a synergistic effect of different probiotic strains under *in vivo* conditions.

## EFFECTS OF PROBIOTICS IN POULTRY

Addition of probiotics has shown beneficial effects on growth performance of poultry as well. In broilers, supplementation of a control diet with probiotics based on *Bacillus cereus* or *Saccharomyces boulardii* improved feed conversion rate by 12 and 11%, respectively (Gil de los Santos *et al.*, 2005). Moreover, after 47 days, average live weight was significantly higher (16 and 7%, respectively) in birds fed the two types of probiotics in comparison to the control group (Figure 6.3).

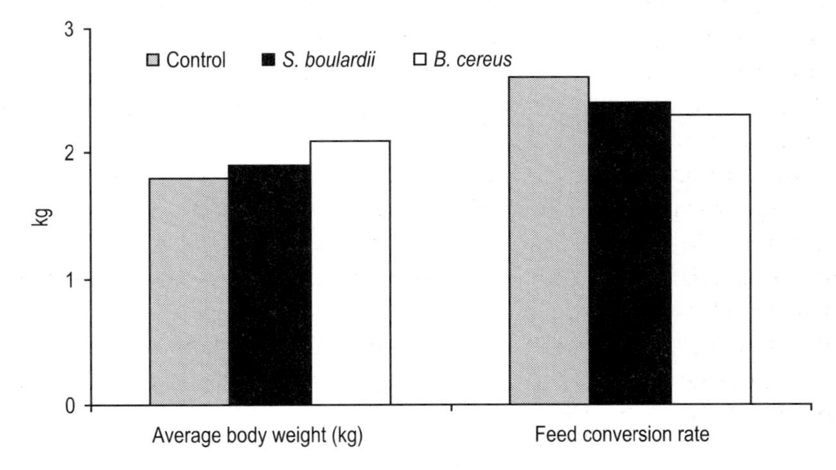


**Figure 6.3.** Average daily live weight and feed conversion rate as influenced by dietary supplementation with different probiotics (adapted from Gil de los Santos *et al.*, 2005)

In turkeys, Männer *et al.* (2002) reported a 1.5 and 2.0% improvement in average daily weight gain and feed conversion rate, respectively, when diets were supplemented with a blend of different probiotic strains originating from *Bacillus* or *Enterococcus*. Furthermore, mortality rate was decreased by 16% in birds fed supplemental probiotics. In a study by Mountzouris *et al.* (2006) with broilers, a multi-strain probiotic additive also increased average daily weight gain and feed efficiency (3 and 2%, respectively). However, these differences were not significant.

## EFFECTS OF PROBIOTICS IN AQUACULTURE

In order to withstand the high stocking densities currently implemented in shrimp hatcheries and grow-out ponds and related stress situations (e.g. low dissolved oxygen contents in the water), probiotics are a promising and sustainable additive to stimulate shrimp growth and secure a low disease response. According to Lückstädt (2006), Massam (2005) and Decamp *et al.* (2005), probiotics represent an effective tool to enhance the survival of shrimp in grow-out and hatchery. Improvement of water quality is another approach of using probiotics in aquaculture. Gram-positive strains such as *Bacillus subtilis* are efficient converters of organic matter. These bacteria, therefore, can contribute to adequate oxygen content in water.

Several trials were performed in Vietnamese hatcheries using a probiotic feed additive with three different strains originating from *Bacillus*, *Enterococcus* and *Lactobacillus*, designed to improve nutrient and energy availability in shrimp larvae. Average results from all farms are shown in Table 6.2.

**Table 6.2. Survival rate (%) in hatcheries using a probiotic blend in shrimp**

| *Provinces/Numbers of hatcheries* | *Control containing antibiotics* | *Probiotic additive* |
|---|---|---|
| 1. province, 15 hatcheries | 50% | 53% |
| 2. province, 12 hatcheries | 45% | 46% |
| 3. province, 9 hatcheries | 55% | 58% |
| 4. province, 20 hatcheries | 50% | 51% |

Since the survival rates of shrimp larvae did not differ ($P > 0.05$) between hatcheries, it can be concluded that the probiotic additive can be an alternative for an antibiotic-free hatchery operation under circumstances found in Vietnam.

## EFFECTS OF PROBIOTICS IN RUMINANT NUTRITION

In ruminants, probiotics have been used to replace AGP in neonatal and stressed calves as well as to increase milk yield in dairy cows and to improve growth performance in beef cattle (Krehbiel *et al.*, 2003). According to Klein (1995), oral administration of a probiotic feed additive based on *Bacillus cereus* var. toyoi improved average daily weight gain and feed efficiency in calves by up to 6 and 5%, respectively. A significant improvement in average daily weight gain (8%) and feed efficiency (12%) was reported by Roth *et al.* (1992b) when calves were fed probiotics based on the same strain of *Bacillus*. Probiotics have also been shown to affect the pattern of ruminal fermentation. Erasmus *et al.* (1992) reported that feeding of a yeast supplement to lactating dairy cows increased the ruminal flow of microbial protein and altered the amino acid composition of this protein fraction.

Moreover, probiotic bacteria are frequently used as silage inoculants in dairy and beef cattle feeding to improve the ensiling process and the aerobic stability of silages. The process of ensiling is based on a lactic acid fermentation under anaerobic conditions that results in a rapid decrease of the pH values in the forage. Although lactic acid-producing bacteria are natural constituents of the epiphytic microbial population in all forages, their initial number is rather low and begins to increase during the first phase of fermentation. Thus, addition of exogenous lactic acid bacteria may contribute to a rapid decline of pH in the forage, which is a main precondition for optimal preservation. Silage inoculants usually based on *Lactobacillus plantarum*, *Enterococcus faecium* or *Pediococcus spp.* may be added in granular, powder or liquid form at dosages ranging between $10^5$ and $10^6$ CFU/g of forage (Weinberg *et al.*, 2003). As reported by Sebastian *et al.* (1996), ensiling of high-moisture ear corn with a mixture of *Lactobacillus plantarurn* and *Enterococcus faecium* significantly increased the lactic acid concentration in comparison to a control without microbial inoculants. A well-selected combination of different microbial strains with different properties (e.g. growth rate, tolerance against low moisture or high acidity, temperature and pH optima) may exert synergistic effects (Kung, 2000). The effects of using probiotic silage additives are supposed to be highest in forages with low contents of carbohydrates and high protein and mineral concentrations.

# 7

# PREBIOTICS

## Characterisation

According to Gibson and Roberfroid (1995), prebiotics are non-digestible feed supplements that beneficially affect the host by selectively stimulating the growth and/or activity of one or a limited number of bacterial species in the digestive tract and thus attempt to improve host health. Thus, prebiotics may enhance the response of digestibility or performance parameters to supplementation with probiotics by creating favourable conditions for beneficial bacteria.

## Required properties of prebiotics

As can be derived from Gibson and Roberfroid (1995), prebiotics must fulfil several criteria in order to exert positive effects on gut health or growth performance. Basically, they must be selectively utilised by beneficial bacteria such as *Lactobacillus*, *Bacteroides*, *Bifidobacteria, Pediococcus* or *Enterococcus*, but not by harmful ones. Thus, it is a prerequisite that prebiotics neither be fermented nor absorbed in the upper part of the gastrointestinal tract.

## Prebiotics used in animal nutrition

According to Bauer *et al.* (2006), several carbohydrates that may be fermented by intestinal microorganisms can be classified as prebiotics, including non-starch polysaccharides (NSP), resistant starch and non-digestible oligosaccharides (NDO). However, among these substrates, inulin and several oligosaccharides (e.g. fructo-oligosaccharides, transgalacto-oligosaccharides or gluco-oligosaccharides) have the highest potential to exert prebiotic effects under practical conditions. Inulin and fructo-oligosaccharides (FOS) are widely used as prebiotic feed additives. Inulin is a polymer consisting of 2-60 fructose units. It is mainly derived from chicory root by extraction. FOS, also referred to as oligofructose, has a lower degree of polymerization, containing up to ten sugar units. Depending on their chemical properties, prebiotics may be fermented selectively by microorganisms in different sections of the gut, including stomach, small intestine and large intestine (Houdijk *et al.*, 1999; Houdijk *et al.*, 1998). Table 7.1 shows the fermentation pattern of several prebiotic carbohydrates for different bacterial strains and species.

**Table 7.1. Bacterial fermentation of different prebiotic carbohydrates (adapted from Hartemink and Rombouds, 1997)**

| *Bacterial group/species* | *Prebiotic carbohydrate*[1] | | | | | | | | |
|---|---|---|---|---|---|---|---|---|---|
| | *FOS* | *INU* | *TOS* | *GLL* | *IMO* | *RAF* | *LAT* | *LAC* | *PHGG* |
| *Bacteroides distasonis* | + | + | + | + | + | +,- | + | + | - |
| *B. fragilis* | + | + | + | + | + | +,- | + | + | - |
| *B. ovatus* | + | + | + | - | + | +,- | + | | + |
| *B. thetaiotaomicron* | + | + | + | | + | +,- | | + | - |
| *B. vulgatus* | + | + | + | + | + | +,- | + | | - |
| *Bifidobacterium spp.* | + | + | + | + | + | + | + | +,- | - |
| *Clostridium butyricum* | | - | - | - | - | | + | + | + |
| *Cl. Clostridioforme* | +,- | - | | - | - | +,- | - | + | - |
| *Cl. Perfringens* | +,- | -,+ | -,+ | - | + | +,- | + | + | - |
| *Cl. Ramosum* | + | + | | - | + | +,- | + | + | |
| *Escherichia coli* | -,+ | - | + | - | - | - | +,- | - | - |
| *Lactobacillus acidophilus-group* | +,- | + | + | + | +,- | - | + | + | - |
| *Lb. casei* | +,- | + | + | - | - | - | + | + | |
| *Mitsuokella multiacidus* | +,- | + | | - | + | + | + | | |
| *Ruminococcus. productus* | - | | | - | - | +,- | + | | |

[1] FOS: fructo-oligosaccharides, INU: inulin, TOS: trans-galactosyl-oligosaccharides, GLL: 4'-galactosyl-lactose, IMO: isomalto-oligosaccharides, RAF: raffinose, LAT: lactulose, LAC: lactitol, PHGG: partially hydrolyzed guar gum

## Mode of action

Due to the lack of suitable gastrointestinal enzymes, prebiotic carbohydrates cannot be digested by nonruminants. Nevertheless, they are exclusively fermented by beneficial bacteria such as *Lactobacilli*, *Bifidobacteria* or *Bacteroides*, therefore having the potential to modulate the composition of microbial communities in the gut. For this reason, they are also referred to as "bifidogenic" feed additives. Microbial degradation of prebiotic carbohydrates results in accumulation of short-chain fatty acids (mainly acetate, propionate and butyrate) and coincides with a lowering of the gut pH, particularly in the large intestine. Moreover, short-chain fatty acids may serve as extra energy, thus contributing to the overall energy supply of the host animal (Muramatsu *et al.*, 1991). Additionally, they have been reported to accelerate the rate of epithelial cell turn-over in the gut (Le Blay *et al.*, 2000). According to a recent report by Zdunczyk *et al.* (2005), supplementation of a control diet with FOS significantly increased the concentrations of short-chain fatty acids (401 vs. 285 μmol/kg body weight) and lowered ileal pH values (5.2 vs. 5.9) in turkeys. In studies by Shim (2005), dietary supplementation of FOS at 3% of finished feed significantly decreased the pH in the caecum (5.4 vs. 5.8) and proximal colon (5.5 vs. 5.9) of piglets. Similar observations were made by Chambers *et al.* (1997) with different prebiotics (FOS, lactose derivates) fed to 5- and 6-week-old broilers. Consequently, the growth of microorganisms that are sensitive to acidic pH conditions is inhibited, therefore protecting the gut from intestinal diseases.

Supporting beneficial bacteria by supplementation of diets for pigs with prebiotics may reduce the level of fermentation of undigested protein by undesirable, proteolytic bacteria in the gut (Mosenthin and Bauer, 2000), which, in turn, results in a lower accumulation of undesirable products such as ammonia, toxic amines, skatole and indole. Consequently, less energy is required for the degradation of such toxic compounds in the liver. In studies by Xu *et al.* (2003), FOS enhanced the growth of *Bifidobacteria* and *Lactobacilli* and inhibited *E. coli* in the small intestine and caeca of broiler chicks (Figure 7.1). In this study, 240 one-day-old broilers were fed a corn-soybean meal-based diet which was supplemented with graded doses of FOS (0, 2, 4 or 8 g/kg of finished feed) at the expense of corn.

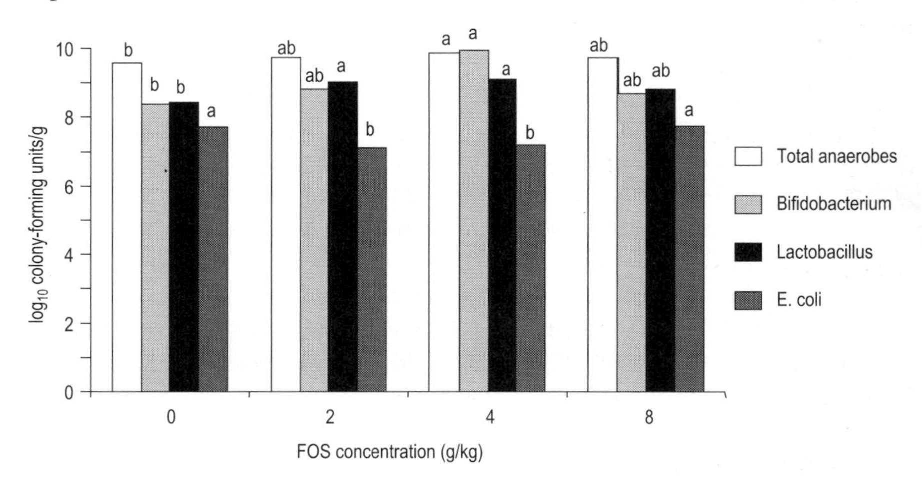


**Figure 7.1.** Effect of graded doses of FOS on microbial counts in caecal digesta of broilers (adapted from Xu *et al.*, 2003)

In the same study, the activity of digestive enzymes (amylase, protease) was enhanced when diets were supplemented with FOS. This observation was attributed to the contribution of digestive enzymes delivered by the increased numbers of *Bifidobacteria* and *Lactobacilli*. Juskiewicz *et al.* (2005) reported that dietary supplementation of inulin significantly reduced the population of *E. coli* in the caeca of turkeys and tended to increase numbers of *Bifidobacteria* and *Lactobacilli*. In contrast, supplementation with a commercial AGP (Flavomycin) tended to decrease the numbers of these beneficial bacteria. Morphological changes in the gut of laying hens were reported by Chen *et al.* (2005) when prebiotics were fed to White Leghorn hens. Supplementation of diets with FOS or inulin at 1% of finished feed significantly increased the length of the small and large intestine of the birds, meaning that the overall absorptive capacity was increased. Consequently, the authors observed a significant increase in gain/feed ratio.

Several studies were directed to investigate the influence of prebiotic supplementation on gut morphology. In studies with weaned piglets, supplementation of wheat-soybean meal diets with FOS numerically increased villus height in different sections of the small intestine (Shim, 2005). In broilers, addition of FOS at 3% of diet significantly increased villus height in the ileum as well as crypt depth in the jejunum and caecum (Xu *et al.*, 2003). As suggested by Xu *et al.* (2003), morphological changes may be attributed to the ability of prebiotics to create a beneficial microbial gut environment rather than to a direct effect of prebiotics on the gut tissue.

The term "synbiotics" describes a mixture of well-selected probiotics and prebiotic carbohydrates, having the potential to exert additive or synergistic effects in terms of gut health, digestibility and performance (Roberfroid, 1998). Nemcova *et al.* (1999) investigated the effect of a probiotic alone or in combination with FOS on the composition of the faecal microflora in piglets. Dietary supplementation of *Lactobacillus paracasei* alone significantly decreased the faecal counts of *Clostridia* and *Enterobacteriaceae*. Moreover, the combined administration of *Lactobacilli* and FOS significantly increased faecal counts of *Lactobacilli*, *Bifidobacteria*, total anaerobes and total aerobes and significantly reduced levels of *Clostridia* and *Enterobacteriaceae*, indicating a strong synergistic effect of probiotics and prebiotics on gut microflora.

In conclusion, supplementation of diets with prebiotics can positively affect the composition of the gut microflora, causing a shift towards non-pathogenic, beneficial species at the expense of harmful species.

## Effects of prebiotics on performance

Reports on the effects of prebiotic feed supplements in terms of digestibility and growth performance are rather equivocal. A lack of response to prebiotic supplementation on digestibility and performance parameters, such as reported by Farnworth *et al.* (1992) or Matthew *et al.* (1997), may be attributed, for example, to excellent experimental conditions, choice of the inclusion level, high initial levels of oligosaccharides in the feed or fermentation of prebiotics by non-beneficial bacteria. The choice of a suitable inclusion level is probably the most important issue. It can be expected that overdosing may cause a depression in growth performance since excessive fermentation of prebiotic carbohydrates may result in anti-nutritional effects such as increased intestinal gas production and diarrhoea (Mosenthin and Bauer, 2000). Positive effects of prebiotic supplementation on performance parameters are presented in the following paragraph.

## EFFECTS OF PREBIOTICS IN PIGS

Orban *et al.* (1997) observed no difference in the efficacies between AGP and prebiotics in piglets fed corn-soybean meal diets. Supplementation with AGP (Apramycin sulphate) or sucrose thermal oligosaccharide caramel (STOC), which is a mixture of fructose-rich oligosaccharides and di-fructose di-anhydrides, increased average daily gain and feed intake by up to 11 and 15%, respectively (Orban *et al.*, 1997). However, these improvements were not significant, presumably due to high variation of these parameters within treatment groups. Similarly, He *et al.* (2002) reported that average daily weight gain of piglets tended to increase by up to 11%, when a control diet based on corn and soybean meal was supplemented with inulin originating from chicory. Similar effects of FOS supplementation on performance parameters were obtained by Russel *et al.* (1996) with weanling pigs. However, the differences obtained in these studies were only numerical. Significant effects of prebiotic addition to diets were reported by Shim (2005). Supplementation of diets for weaned pigs with FOS significantly increased apparent total tract digestibility of crude protein and minerals by 7 and up to 5%, respectively. Consequently, average daily gain and feed efficiency were increased by 19 ($P < 0.05$) and 8% ($P > 0.05$), respectively.

## EFFECTS OF PREBIOTICS IN POULTRY

In a report by Xu *et al.* (2003), supplementation of broiler diets with FOS at 0.4% of finished feed significantly improved average daily gain and significantly reduced feed conversion ratio by 11 and 9%, respectively.

However, feed intake was not affected by FOS supplementation in this study. In another trial, supplementation of a control diet with FOS (0.4%) significantly improved average daily gain by 10% (Zhan *et al.*, 2003). Furthermore, feed/gain ratio was significantly reduced by 5 and 8%, when FOS was included at 0.2 and 0.4%, respectively. However, supplementation of FOS at higher dosage (0.6%) did not affect performance parameters in the trial by Zhan *et al.* (2003). In a study by Chen *et al.* (2005), addition of FOS or inulin (1%) significantly increased weekly egg production by 13 and 11%, respectively. However, there were no differences in average egg weight or daily feed intake, but feed efficiency was significantly improved. The impact of a synbiotic preparation containing as major ingredients *Enterococcus faecium* and FOS on growth performance was investigated in a study with 200 one-day-old broilers. Administration of the synbiotic in the drinking water significantly improved live weight (5.5%) and daily weight gain (5.5%) and numerically reduced feed/gain ratio (5.9%) and mortality rate (57.7%) (Figure 7.2).

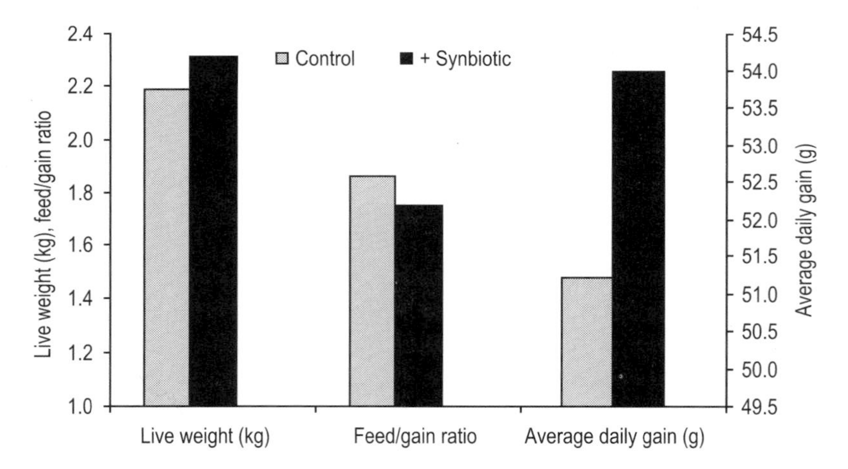


**Figure 7.2.** Effect of synbiotic additives, NGPs™ (Biomin® IMBO), on performance parameters in broilers (Mohnl *et al.*, 2006). NGPs™ is an applied trademark of Erber AG.

Moreover, synbiotic supplementation increased the Broiler Productivity Index, i.e. [liveability × live weight] / [age × feed/gain ratio] × 100, by 18.4%.

## EFFECTS OF PREBIOTICS IN AQUACULTURE

In aquatic species, several carbohydrates such as FOS (Panigrahi *et al.*, 2005) or ß-glucans (Wang and Wang, 1997) have been used, however, primarily for the purpose of immune prophylaxis. As will be discussed in Chapter 10, these carbohydrates have also the potential to increase survival rate and growth performance in several fish species.

## EFFECTS OF PREBIOTICS IN RUMINANTS

AGP have been extensively used in the feeding of calves to protect them from pathogenic invasion in the early stages of life. Few studies have been carried out to examine the effect of fermentable carbohydrates as replacements to AGP in calves (e.g. Quigley *et al.*, 2002; Quigley *et al.*, 1997; Webb *et al.*, 1992). Quigley *et al.* (1997) investigated the effect of supplementation of a milk replacer with an oligosaccharide (galactosyl-lactose) in comparison to supplementation with AGP in bull calves. Galactosyl-lactose was more efficient than the AGP resulting in an improvement in average daily weight gain by 58%, whereas the antibiotic increased weight gain by 42% only.

# 8

# FEED ENZYMES

## Characterisation

Enzymes are used on a large scale in various industries, including, among others, milk processing, alcohol production, cosmetics, diagnostics, textiles, paper, leather and detergents. Enzymes are bio-catalytic proteins that initiate or accelerate specific biological reactions. The use of enzymes in animal nutrition has evolved since the early 1980s and the worldwide demand for feed enzymes is supposed to grow by double-digit gains, especially in the Asian-Pacific area.

In general, feed enzymes are capable of degrading specific substrates by cleaving defined bonds in the respective molecule. Thus, they can be classified according to the type of the substrate on which they act; i.e.:

- Substrates such as starch, proteins, lipids, which are also degraded by endogenous enzymes of the host (i.e. amylases, proteases, lipases)
- Substrates such as fibre components (e.g. cellulose) including substrates with anti-nutritional properties (e.g. phytate, pentosans), which are not degraded by digestive enzymes of the host,

Due to the strict substrate specificity, the choice of suitable enzymes for application in animal nutrition has to be based on the ingredient composition of diets. The most important groups of feed enzymes used in animal nutrition include non-starch polysaccharides-degrading enzymes and phytases. Furthermore, amylases and proteases may be successfully used under certain conditions in order to support the host's endogenous enzyme secretion.

The efficacy of enzymes, in general, depends on several factors such as surrounding pH and temperature. Enzymes from different sources may considerably differ in their optima with regard to these parameters. Thus, determination of enzyme activities is highly complex and a direct comparison of products purely based on the activities measured is not always feasible. One example is the determination of phytase activity, generally carried out at pH 5.5, which is the pH optimum of *Aspergillus niger* phytase, according to the procedure outlined by Engelen *et al.* (1994). However, *E. coli* phytase, for example, has a pH optimum between 2.5 and 3.5 (Rodriguez *et al.*,

1999). Therefore, measurements of efficacy of these two phytases are confounded by differences in pH optima.

## Non-starch polysaccharides-degrading enzymes

In Europe and many other regions worldwide, cereals and oilseed meals, representing valuable sources of energy and protein, are commonly used as ingredients in animal feeds. However, these feedstuffs contain considerable amounts of indigestible carbohydrates as well. These fractions of carbohydrates are also referred to as non-starch polysaccharides (NSP). As shown in Figure 8.1, NSP can be classified as cellulose, non-cellulosic polymers (hemicellulose) and pectic polysaccharides (Bailey, 1973).

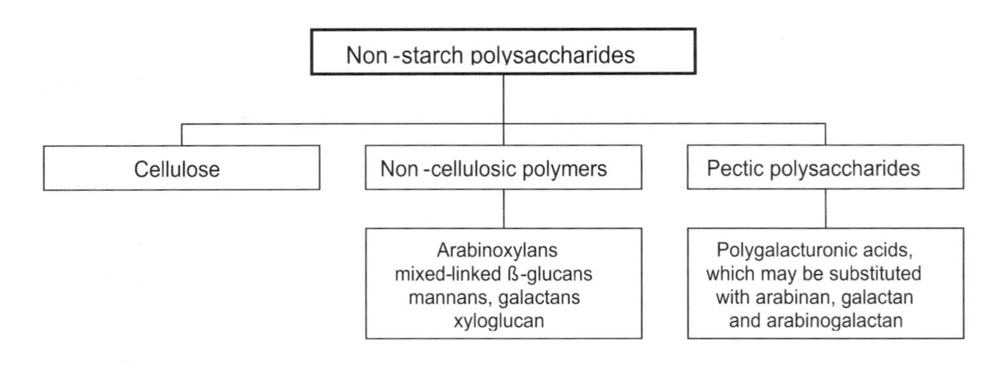


**Figure 8.1.** Classification of non-starch polysaccharides

A number of soluble and insoluble NSP have been identified in cereals and oilseed meals (Table 8.1), among which the most important are arabinoxylans, ß-glucans and cellulose. Soluble and insoluble NSP may exert different antinutritional effects in nearly all domestic animals. Soluble NSP have the ability to bind water, thereby increasing viscosity of the digesta in the gut. Insoluble NSP are known to encapsulate nutrients (e.g. starch, protein or lipids) within the fibre matrix of cell walls. Thus, insoluble NSP represent a physical barrier between these nutrients and digestive enzymes of the host, a phenomenon also named the "cage effect". Moreover, it has been suggested that the incidence of gastrointestinal disorders such as dysentery or post-weaning colibacillosis may be associated with high dietary levels of NSP (Hopwood *et al.*, 2004; Pluske *et al.*, 1996).

**Table 8.1. NSP contents (% dry matter) in different cereals and oilseed meals (adapted from Probert, 2004; Bach Knudsen, 1997; Choct, 1997 and Englyst *et al.*, 1989)**

| | *Arabinoxylans* | *ß-glucans* | *Cellulose* | *Total NSP* |
|---|---|---|---|---|
| Wheat | 8.1 | 0.8 | 2.0 | 11.4 |
| Rye | 8.9 | 2.0 | 1.5 | 13.2 |
| Triticale | 10.8 | 1.7 | 2.5 | 16.3 |
| Barley | 7.9 | 4.3 | 3.9 | 16.7 |
| Oats | 15.5 | 2.8 | 10.1 | 23.3 |
| Corn | 5.2 | - | 2.0 | 8.1 |
| Soybean meal | 3.7 | - | 4.0 | 17.0 |
| Rapeseed meal | 6.5 | - | 6.6 | 20.0 |
| Sunflower meal | 8.7 | - | 11.4 | 26.0 |

Barley, oats, rye, triticale and wheat contain significant amounts of soluble NSP. Inclusion of these "viscous" grains in diets results in sticky droppings and poor litter quality, predominantly in poultry. Thus, enzymes specifically tailored to decrease gut viscosity are recommended in diets for broilers, layers and turkeys. In pigs, the negative impact on digestibility, performance and litter parameters resulting from the viscous properties of soluble NSP seems to be less important as compared to poultry (Partridge, 2001). This can be attributed to a longer retention time and a higher number of microorganisms inhabiting the gastrointestinal tract of pigs. After passage through the small intestine, large quantities of NSP are degraded by the intestinal microflora in the hindgut. This results in increased production of volatile fatty acids (Montagne *et al.*, 2003), which, in turn, decreases the faecal pH values (Canh *et al.*, 1998).

Since in young piglets, the gut is rather immature and the gastrointestinal secretion of digestive enzymes is not yet fully developed, Diebold (2005) concluded that supplementation of starter and pre-starter diets with NSP-degrading enzymes has a higher impact on digestibility and performance parameters in piglets than in growing/finishing pigs or sows.

In some cases, the effect of NSP-degrading enzymes on digestibility and performance parameters produced inconsistent results. Inborr *et al.* (1993), for example, observed no effect on growth performance when xylanase in combination with amylase and ß-glucanase was added to wheat-barley-based diets for early-weaned piglets. In another study performed with weaned piglets (Diebold *et al.*, 2004), dietary supplementation with xylanase did not affect the apparent ileal digestibilities of organic matter, energy, fat, crude protein and amino acids. The lack of response to enzyme supplementation might be attributed to the high complexity of NSP in plant feedstuffs, which requires a mixture of specific enzymes to degrade these substrates. Thus, addition of multi-enzyme preparations to diets is recommended in order to optimise the degradation of different NSP in the gut. The use of specific NSP-degrading enzymes allows for inclusion of lower-quality feedstuffs into diets, thus potentially reducing the feed costs while maintaining performance.

Feed enzymes are frequently added to diets in powder or granular form. However, being high-molecular proteins, enzymes are sensitive to heat. During feed processing (e.g. pelleting, extrusion, expanding), temperatures may exceed 80 °C, which is detrimental to feed enzymes. This requires post-pelleting spray application of liquid enzyme preparations which may be rather expensive for feed manufacturers. As an alternative, there have been recent attempts to apply feed enzymes via the drinking water (Maisonnier-Grenier *et al.*, 2005).

In general, feed enzymes are industrially produced from fungi, bacteria or yeasts by fermentation under defined conditions. NSP-degrading enzyme preparations available on the market contain at least one of the following activities:

- xylanase (endo-1,4-ß-xylanase, EC 3.2.1.8)
- ß-glucanase (endo-1,3(4)-ß-glucanase, EC 3.2.1.6)
- cellulase (endo-1,4-ß-glucanase, EC 3.2.1.4)
- α-galactosidase (EC 3.2.1.22)
- pectinase (polygalactonurase, EC 3.2.1.15)

However, since purification of enzymes is rather expensive and time-consuming, most products contain, apart from one major activity, different side activities. These side activities, however, are rather variable, whereas the major activity must be clearly defined.

## MODE OF ACTION OF NSP-DEGRADING ENZYMES

As illustrated in Figure 8.2, the mode of action of NSP-degrading enzymes is highly complex. There are several effects which may contribute to the overall outcome of enzyme supplementation, including a decrease in digesta viscosity, softening of the cage effect, modifications of the gut microflora as well as morphological and physiological modification of the gut.

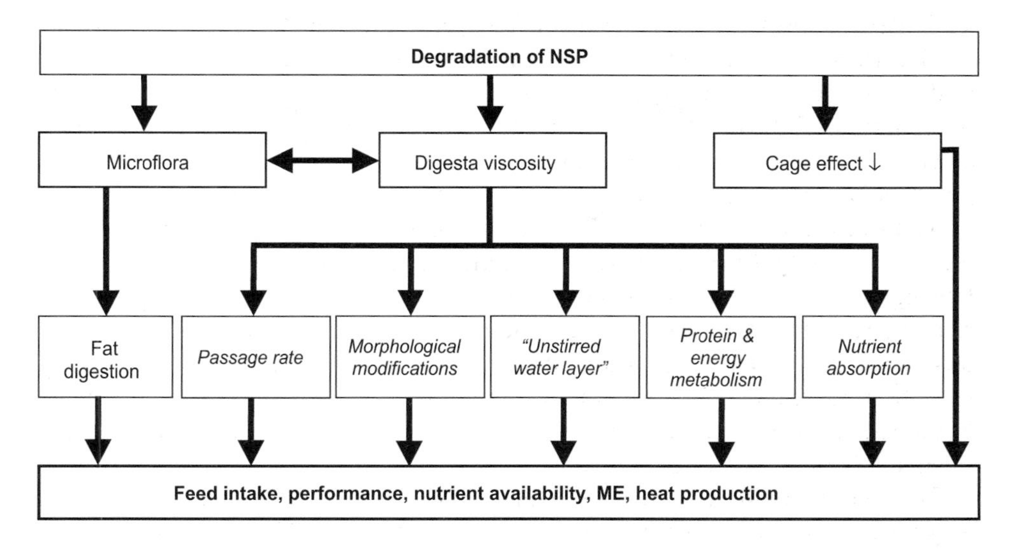


**Figure 8.2.** Mode of action of NSP-degrading enzymes (Simon, 1998)

### *Effects of NSP-degrading enzymes on the fraction of insoluble NSP*

Supplementation of diets containing high amounts of insoluble NSP with suitable NSP-degrading enzymes results in a softening of the "cage effect". Thus, enzyme supplementation has the potential to facilitate the accessibility of endogenous digestive enzymes and substrates, resulting in improved availability of nutrients and energy for the animal (Yin *et al.*, 2001a; Simon, 1998). In weaned piglets, supplementation of diets based on hulless barley with xylanase and ß-glucanase significantly increased apparent ileal digestibility of NSP (xylose, arabinose, galactose) (Yin *et al.*, 2001a). In studies with broilers, addition of NSP-degrading enzymes (xylanase, ß-glucanase, cellulase, pectinase, mannanase, galactanase and combinations thereof) to wheat-based diets significantly reduced the concentration of insoluble NSP in the digesta of broilers, thereby improving ileal digestibilities of NSP, fat, starch and nitrogen (Meng *et al.*, 2005; Meng *et al.*, 2004).

### *Effects of NSP-degrading enzymes on digesta viscosity*

Supplementation of cereal-based diets with NSP-degrading enzymes which target the soluble NSP fraction results in a decrease in digesta viscosity and water-holding capacity (Dänicke, 1999). Consequently, the transit of digesta in the gut is accelerated, which, in turn, may stimulate voluntary feed intake. It is assumed that a reduction in digesta viscosity improves the rate of diffusion between host digestive enzymes (proteases, lipases, amylases) and nutrients, resulting in increased nutrient digestibility and energy availability. As reported by Meng *et al.* (2004), supplementation of control diets with an enzyme cocktail containing ß-glucanase, xylanase and cellulase significantly reduced digesta viscosity in the jejunum of broilers. Consequently, ileal nutrient digestibilities in conjunction with average daily gain and feed efficiency were significantly improved. Similar results were reported for pigs. In studies by Yin *et al.* (2001a, b), dietary supplementation with ß-glucanase and xylanase significantly decreased digesta viscosity in the distal ileum and significantly increased ileal digestibility of energy and most amino acids.

***Effects of NSP-degrading enzymes on gut microflora***

Microorganisms colonising the gut are dependent on the supply of nutrients originating from the feed. Thus, the gut microflora, both in terms of quantitative and qualitative aspects, is highly affected by the composition of the feed. Highly-digestible diets based on corn and animal protein sources (e.g. fish meal) provide only little substrate for gut microorganisms since most of the nutrients are released and absorbed in the small intestine. In contrast, the use of cereals and oilseed meals increases the dietary content of poorly-digestible NSP, thus leaving more substrate for gut microorganisms. Furthermore, the impact of NSP on digesta viscosity may affect the composition of the gut microflora as well. An increase in viscosity generally promotes bacteria with longer life time cycles (Simon, 1998).

The use of NSP-degrading enzymes may assist in protecting animals from pathogenic invasion (Choct, 2004). As reviewed by Bedford and Apajalahti (2001), dietary supplementation with NSP-degrading enzymes has the potential to reduce the total number of microorganisms in the gut. This can be attributed to an increase in ileal nutrient digestibility in conjunction with decreased supply of available nutrients for the gut microorganisms. It is also assumed that administration of NSP-degrading enzymes may exert secondary, prebiotic effects on the gut microflora since the partial degradation and solubilisation of insoluble NSP through suitable enzymes results in sugars and specific oligomers (e.g. xylo-oligosaccharides) which may selectively promote the growth of beneficial bacteria (Vahjen *et al.*, 1998). Sinlae and Choct (2000), for example, reported that addition of xylanase to wheat-based diets reduced the total number of *Clostridia* in the gut of broilers.

Dietary fat represents an important source of energy in livestock feeds. As speculated by Knarreborg *et al.* (2003), specific gut microorganisms (e.g. *Lactobacilli* and *Clostridia*) may impair digestion of lipids by production of enzymes that contribute to the deconjugation of bile acids in the gut. Thus, any alteration of the gut microflora caused by addition of supplemental feed enzymes will probably have an impact on digestion of dietary fat and also on the utilisation of fat-soluble vitamins (Simon, 1998) such as vitamin A or D.

***Effects of NSP-degrading enzymes on gut morphology***

Addition of feed enzymes may also induce morphological and structural changes in the gut tissues (Ikegami *et al.*, 1990). It can be derived from Simon (1998), that increasing dietary levels of "viscous" cereals cause an increase in weight of the total gut, indicating a close relationship between digesta viscosity and gut morphology. Furthermore, several authors reported that high NSP concentrations might stimulate the development of goblet cells on the epithelial gut surface (Viveros *et al.*, 1994; Schneemann *et al.*, 1982). Simon (1998) reported that the intensity of mucin secretion, as affected by addition of NSP-degrading enzymes, may influence the thickness of the so-called "unstirred water layer" which has a significant impact on nutrient absorption and utilisation.

Ritz *et al.* (1995) indicated that addition of exogenous enzymes might improve nutrient digestibility by increasing the overall absorptive surface area of the gut. In studies with turkeys, supplementation of control diets with amylase increased the length of the villi in the jejunum and ileum and improved feed efficiency and average daily gain in the first two or three weeks of the experiment, respectively. However, addition of xylanase to the diets did not affect the intestinal morphology or growth performance parameters in the study by Ritz *et al.* (1995). Yasar and Forbes (2000) reported that dietary supplementation with feed enzymes (xylanase, protease and ß-glucanase) significantly reduced the relative gut weight (expressed as g/kg of body weight), but had only little effect on the thickness of the gut mucosa.

***Further effects of NSP-degrading enzymes***

Legume seeds and oilseed meals are commonly incorporated in animal feeds as sources of protein and amino acids. However, the NSP in these feedstuffs are generally more complex as compared to cereals. Furthermore, they contain considerable amounts of antinutritional factors, including different oligosaccharides such as ß-galactomannans or $\alpha$-galactosides that may cause flatulence and diarrhoea, predominantly in young nonruminants (Rackis, 1981). Therefore, specific enzymes tailored to degrade these antinutritional compounds may beneficially affect nutrient digestibility and growth performance in young pigs or poultry fed diets with high contents of legume seeds or oilseed meals. In studies by Kim *et al.* (2003), supplementation of a basal diet containing high levels of soybean meal (20-32%) with an enzyme blend mainly containing α-galactosidase, ß-mannanase and ß-mannosidase significantly improved overall gain/feed ratio by up to 9% in nursery pigs. Additionally, a significant improvement (11%) in weight gain and apparent ileal digestibility of some amino acids (e.g. lysine, threonine, tryptophan, histidine) was obtained in the third week of a second trial with nursery pigs. Similarly, in a report by Pettey *et al.* (2002), addition of ß-mannanase to corn-soybean meal-based diets significantly improved average daily weight gain in finishing barrows. However, there were no effects of enzyme supplementation on apparent total tract digestibility of energy, dry matter, nitrogen and phosphorus.

## EFFECTS OF NSP-DEGRADING ENZYMES ON PERFORMANCE

Supplementation of diets with exogenous NSP-hydrolysing enzymes results in degradation of NSP to carbohydrates with a lower degree of polymerisation, thereby reducing the negative impact of high digesta viscosities and cage effect. In consequence, nutrient digestibility and energy availability are improved, which, in turn, may result in increased body weight gain and gain/feed ratio.

### *Effects of NSP-degrading enzymes in pigs*

In studies with piglets, Yin *et al.* (2001b) obtained an increase in ileal digestibilities of dry matter, gross energy, crude protein, fibre (Figure 8.3) and some amino acids (e.g. lysine, threonine, glycine) when diets based on hulless barley were supplemented with a xylanase, ß-glucanase or an enzyme cocktail.

Xylanases, ß-glucanases or cellulases may be used in combination with endogenous enzymes (e.g. amylase, protease) in order to support the endogenous digestive system of young animals. Omogbenigun *et al.* (2004) investigated the efficacy of an enzyme cocktail containing xylanase, ß-glucanase, amylase, protease, invertase and phytase in weaned piglets fed wheat-based diets. In this study, enzyme addition significantly improved average daily gain and gain/feed ratio by up to 18 and 17%, respectively.

Similar effects were reported by Dusel *et al.* (1997) and Jeroch *et al.* (1999) in studies with piglets fed different NSP-degrading enzymes.

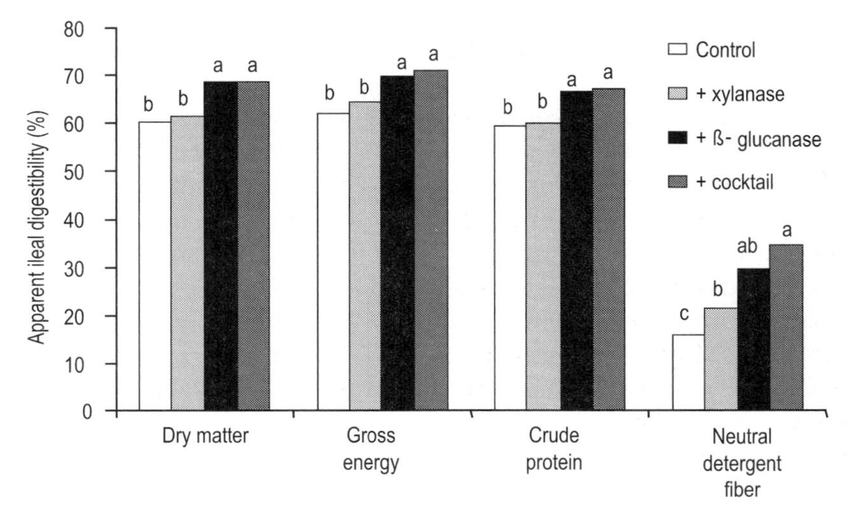


**Figure 8.3.** Apparent ileal digestibility of dry matter, gross energy, crude protein and neutral detergent fibres as influenced by different types of enzyme supplementation (Yin *et al.*, 2001b)

### *Effects of NSP-degrading enzymes in poultry*

Beneficial effects of NSP-degrading enzymes in poultry nutrition are mainly due to their impact on digesta viscosity. A positive influence on performance parameters has been reported by Zhang *et al.* (1997) in broilers fed rye-based diets supplemented with different xylanase preparations at different dosages. Similarly, in studies with broilers (Choct *et al.*, 2004), supplementation of wheat-based diets with xylanase preparations originating from different fungal sources improved average daily weight gain and feed conversion rate by 4-14 and 4-9%, respectively. However, the magnitude of response of these parameters clearly depended on the origin of the added enzymes, indicating differences in substrate specificity between different sources of NSP-degrading enzymes. Meng *et al.* (2004) investigated the efficacy of NSP-degrading enzymes in broilers. In this study, supplementation of wheat-based diets with xylanase, ß-glucanase and cellulase significantly improved average daily gain by 5% and lowered feed/gain ratio by 3%. Moreover, enzyme supplementation tended to increase feed intake in the study of Meng *et al.* (2004).

### *Effects of NSP-degrading enzymes in aquaculture*

The production of carnivorous aquatic species (e.g. salmon, shrimp) is supported by a large input of fish meal or fish oil to the farming system, mainly from ocean fisheries. New and Csavas (1995) asked: "Will there be enough fish meal for fish meals?" Almost one third of the total world fish harvest (142 million tonnes) in the year 2000 was processed to fish meal or fish oil for use as animal feeds. Therefore, aquaculture competes with other livestock production sectors for the limited fish meal resources. It is, therefore, of great interest to replace fish meal by plant protein sources (Mabahinzireki *et al.*, 2001). Soybean meal, for example, is a valuable plant protein source and is widely available as feed ingredient. However, its use in aquaculture is limited due to the presence of several antinutritive factors such as protease inhibitors, tannins, lectins or α-galactosides which may have a negative impact on amino acid digestibility and growth performance (Pongmaneerat and Watanabe, 1992a, b). The use of exogenous enzymes in feed for aquatic species is expected to increase rapidly within the forthcoming years (Hardy, 2000). However, studies investigating the efficacy of feed enzymes in fish or shrimp are scarce. Ogunkoya *et al.* (2005) examined the effect of an enzyme cocktail (xylanase, amylase, cellulase, protease and ß-glucanase) and increasing levels of soybean meal in rainbow trout fed diets based on fish meal. However, besides a significant reduction in cohesiveness of the faeces, enzyme addition exerted no effect on growth performance and only marginal and inconsistent effects on apparent digestibility coefficients. The lack of efficacy in this study may be attributed to the fact that the basal diets were already highly digestible. Apparent digestibility of crude protein, for example, ranged between 92 and 94% in all treatments. Further development of feed enzymes and application technologies in the aquaculture sector is mandatory in the next few years.

### *Effects of NSP-degrading enzymes in ruminants*

Due to the activity of ruminal microorganisms, NSP are generally well digested in ruminant species. However, several studies were carried out to investigate the potential use of exogenous NSP-degrading enzymes in ruminant nutrition. A comprehensive review has previously been published by Beauchemin *et al.* (2004). As reported by Yang *et al.* (1999) and Feng *et al.* (1996), addition of cellulase or hemicellulase successfully improved digestibility of the fibre fraction in cattle. Furthermore, Beauchemin *et al.* (2004) indicated that supplementation of diets with NSP-degrading enzymes may increase milk production in dairy cows and weight gain in beef cattle. In a study by Kung *et al.* (2000) with dairy cows fed a total mixed ration (50% forage, 50% concentrate), an enzyme preparation based on cellulase and xylanase significantly increased milk yield when compared to a control treatment without enzyme addition (39.5 vs. 37.0 kg). However, the response of performance parameters to enzyme supplementation is rather variable in ruminants. Dhiman *et al.* (2002), for example, obtained no differences in feed intake, milk yield, milk composition or body weight gain, when cows were fed total mixed rations with or without supplemental NSP-degrading enzymes. Various factors may be responsible for differences in efficacy of exogenous enzymes in ruminants. Due to more suitable moisture and pH conditions, for example, the efficacy of enzyme addition is supposed to be higher in silages as compared to dry feeds (Beauchemin *et al.*, 2004).

## Phytases

The main purpose of phytase supplementation in nonruminant nutrition is to improve phosphorus digestibility. Therefore, phytase enzymes do not necessarily belong to the category of NGP in the strictest sense. However, increased growth performance has been reported under certain conditions when phytase has been added to the feed. Phytases have proven to substantially reduce phosphorus emissions from livestock production systems and have gained considerable importance in modern animal nutrition.

The major portion (50-80%) of total phosphorus in plant feedstuffs is present as phytate (Figure 8.4), the poorly-soluble salt of phytic acid (Lantzsch, 1990). Moreover, phytate may bind mineral cations (e.g. calcium, zinc, iron), protein, amino acids and starch, thereby reducing their bioavailability as well (Maga, 1982; Erdmann, 1979). Moreover, phytate may inhibit protein digestibility by forming complexes with digestive enzymes. In the form of phytate, phosphorus and other bound nutrients are unavailable for nonruminants such as pigs, poultry and fish, because these species lack intrinsic gastrointestinal phytases (Pointillart *et al.*, 1984).

Consequently, high levels of phosphorus are excreted from intensive livestock production systems, thus creating environmental problems associated with the phenomenon of eutrophication (Daniel *et al.*, 1998). Therefore, phytase has been developed as an “environmental” enzyme to alleviate the negative impact of excessive phosphorus excretion in the manure (Kies *et al.*, 2001). Phytase hydrolyzes the phytic acid molecule in the digestive tract and provides inorganic phosphorus and other released nutrients for absorption in the small intestine. In consequence, there is a great reduction in phosphorus excretion when phytase is added to diets for nonruminants.

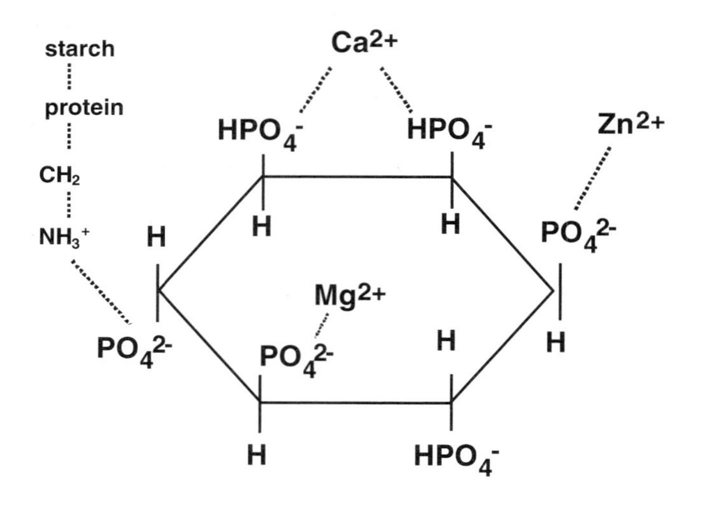


**Figure 8.4**. Structure of phytate (modified after Kemme, 1998)

## OCCURRENCE

Phytases (EC 3.1.3.8) are present in microorganisms such as fungi, bacteria or yeasts as well as in plants. The most commonly used phytase preparations are produced from fungi, for example *Aspergillus niger* or *Peniophora lycii*. At present, phytase derived from *E. coli* has been receiving growing attention. Cereals and cereal by-products contain substantial native phytase (EC 3.1.3.26) activities, whereas oilseed meals and legume seeds exhibit low phytase activities (Steiner *et al*., 2006c; Eeckhout and De Paepe, 1994). However, plant phytases play no role in the feed industry because activities are largely variable and the enzymes are less efficient as compared to microbial phytases (Steiner *et al*., 2006b; Zimmermann *et al*., 2002).

## MODE OF ACTION

Phytase catalyzes stepwise dephosphorylation of phytate in the stomach of nonruminants, thereby producing *myo*-inositol and orthophosphate as end products (Maga, 1982). Thus, addition of phytase to plant-based diets substantially improves phosphorus digestibility in pigs, poultry and fish. Additionally, degradation of phytate may improve the bioavailability of minerals, starch, protein and amino acids as well (Kies *et al*., 2001).

## EFFECTS OF PHYTASE ON DIGESTIBILITY AND PERFORMANCE

The use of microbial phytase for improving phosphorus digestibility in nonruminants has been intensively studied during the last decades. Addition of phytase to the feed substantially reduces phosphorus excretion in the manure by up to 50% in different species (van Heugten and van Kempen, 2001). Most research on phytase has been carried out in pigs and poultry. Although the major effect of phytase supplementation is a considerable increase in phosphorus digestibility, some researchers obtained positive effects on performance parameters as well. The positive impact of phytase supplementation on performance may be attributed to increased protein and energy availability (Kies *et al.*, 2001).

### *Effects of phytase in pigs*

Apparent phosphorus digestibility was increased by 20% through supplementation with 932 U/kg microbial phytase in growing pigs (BW 40 kg) (Gebert *et al.*, 1999). Similar results were obtained in piglets (Kornegay and Qian, 1996). Apparent phosphorus digestibility was increased by 12 and 19 percentage points when microbial phytase was included at levels of 350 and 700 U/kg of feed, respectively. As reported by Jongbloed *et al.* (1996a), phytase supplementation may also increase growth performance in pigs. Based on data from eleven experiments, the authors calculated that feed intake and growth rate were increased by 6 and 3%, respectively, when phytase was supplemented to diets not limiting in dietary phosphorus. Moreover, feed conversion ratio (corrected for differences in weight gain) was decreased by 4.3%.

### *Effects of phytase in poultry*

Phytase supplementation reduces the phosphorus contents of poultry litter. In broilers, supplementation of corn-soybean meal diets with microbial phytase at 500 or 1000 U/kg feed significantly increased apparent ileal digestibility of phosphorus and some amino acids (tryptophane, valine) and also improved feed intake and feed efficiency (Dilger *et al.*, 2004). In laying hens, phytase addition to corn- or barley-based diets significantly increased egg production, feed intake and average weight gain of the hens (Francesch *et al.*, 2005). According to Kies and Schutte (1997), phytase supplementation between 500 and 1000 U/kg feed significantly improved average weight gain and feed efficiency by 1-2%.

***Effects of phytase in aquaculture***

Since around 70% of the total phosphorus sources in plant ingredients are bound as phytate and therefore practically not available for fish (Lall, 1991) the use of phytase came into consideration as well. Liebert and Portz (2005) showed a growth improving effect by microbial phytase in levels between 750 and 1250 U/kg in Nile tilapia, which went in line with a significantly increased energy, protein and phosphorous utilisation as well. Similar effects have been reported for instance in rainbow trout as well (Sugiura *et al.*, 2001).

***Effects of phytase in ruminants***

Since a large portion of dietary phytate is degraded by phytases originating from indigenous ruminal microorganisms, ruminants usually do not rely on supplemental exogenous phytase (Shanklin, 2001). However, Matsui (2002) indicated that phytate phosphorus utilisation by rumen microorganisms is actually not complete. Thus, theoretically, there is a potential to use phytase in feed for ruminants. Hence, Kincaid *et al.* (2005) indicated that exogenous phytase might increase ruminal phytate hydrolysis in lactating Holstein cows to a certain extent. However, phytase addition to total mixed rations only tended to increase phosphorus digestibility in the study of Kincaid *et al.* (2005). In lambs, phytase addition did not exert any effect on phosphorus absorption, although the concentration of phosphorus in the ruminal fluid was significantly increased when phytase was added to the feed (Shanklin, 2001).

# 9

# PHYTOGENICS

## Characterisation

Nearly all plants can be expected to protect themselves from diseases and animal herbivores by accumulation of secondary metabolites in their tissues. Growing concern about AGP in animal nutrition has created efforts to use different plant compounds as possible natural alternatives. Phytogenic feed additives (phytogenics) are an extremely heterogeneous group of feed additives originating from leaves, roots, tubers or fruits of herbs, spices or other plants. They are either available in a solid, dried and ground form or as extracts or essential oils. Phytogenic feed additives usually vary seriously in their chemical ingredients, depending on their composition and influences of climatic conditions, locations or harvest time. Hence, differences in efficacy between phytogenic products which are currently available on the market can be attributed mainly to differences in their chemical composition.

## Herbs, spices and essential oils used in animal nutrition

There is a large number of herbs and spices that may be considered as NGP in animal nutrition, of which the most frequently used are presented in Table 9.1. Most of these plants contain a considerable number of active substances, which determines their *in vivo* efficacy.

**Table 9.1. Herbs and spices frequently used in phytogenic feed additives**

| *Herb/spice* | *Latin name* | *Plant family* | *Main constituents* |
|---|---|---|---|
| Oregano | *Oreganum vulgare* | Labiateae | Carvacrol, thymol |
| Thyme | *Thymus vulgare* | Labiateae | Thymol, carvacrol |
| Garlic | *Allium sativum* L. | Alliaceae, Liliaceae | Diallyldisulphide, allin, allicin |
| Horseradish | *Armoracia rusticana* | Brassicaceae | Allyl-isothiocyanate |
| Chilli, Cayenne Pepper | *Capsicum frutescens* | Solanaceae | Capsaicin |
| Peppermint | *Mentha piperita* | Labiateae | Menthol, carvacrol |
| Cinnamon | *Cinnamomum cassia* | Lauraceae | Cinnamaldehyde |
| Anise | *Pimpinella anisum* | Apiaceae, Umbelliferae | Anethol |

Among the herbs and spices listed in Table 9.1, oregano is probably used most frequently. It is rich in carvacrol and thymol, which are known to have strong antibacterial and antioxidative properties. According to Kamel (2000), oregano and cinnamon showed broad *in vitro* efficacy against various pathogenic bacteria including *E. coli*, *Salmonella typhimurium* and *Clostridium perfringens*.

Essential oils are odoriferous, secondary plant products which contain most of the plant's active substances, being mainly hydrocarbons (e.g. terpenes, sesquiterpenes), oxygenated compounds (e.g. alcohol, aldehydes, ketones) and a small percentage of non-volatile residues (e.g. paraffin, wax) (Losa, 2000). They are obtained from the raw materials, basically through steam distillation. The use of essential oils has a long tradition, e.g. in perfumes, food flavours, deodorants or pharmaceuticals. In human nutrition, their appetising and digestion-promoting effects have long been recognised and aromatic plants are traditionally used in human and veterinary medicine. Their chemical composition is highly diverse, including a relatively large number of unidentified substances. Some biologically active compounds found in essential oils of herbs and spices are shown in Table 9.2.

**Table 9.2. Numbers of different biologically active compounds with different effects present in herbs and spices (adapted from Mathe, 1996)**

| *Herb/spice* | *Effect* | | | | |
|---|---|---|---|---|---|
| | *Antioxidant* | *Sedative* | *Antidepressant* | *Antiviral* | *Bactericide* |
| Bay | 3 | 5 | - | 5 | 5 |
| Cassia | 3 | - | - | 3 | 3 |
| Cayenne | 9 | 7 | 7 | 6 | 8 |
| Cumin | 5 | 6 | - | 7 | 11 |
| Garlic | 9 | 5 | 5 | 5 | 13 |
| Ginger | 6 | 11 | 5 | 6 | 17 |
| Oregano | 14 | - | - | 11 | 19 |
| Rosemary | 12 | 6 | - | 10 | 19 |
| Sage | 7 | - | - | | 6 |
| Thyme | 4 | - | 3 | 3 | 5 |

As shown in Table 9.2, oregano essential oils contain 14 chemical substances with antioxidant properties. Furthermore, as many as 19 substances have been identified to exert bactericidal effects.

## Mode of action

Due to their aromatic properties, several phytogenic feed additives have a positive impact on feed palatability, depending, however, on the applied dosage of the respective ingredients. Their potential to stimulate the voluntary feed intake, especially in young animals, has been reported in several studies (Ertas *et al.*, 2005; Giannenas *et al.*, 2003; Günther and Bossow, 1998). At present, relatively little is known about the mode of action of phytogenics, which may exert beneficial effects on gut health and growth performance. Due to the large number of active compounds, it is rather difficult to attribute these effects to a specific ingredient. Therefore, it is a major aim of current research to elucidate the mode of action of phytogenics in order to establish their routine use in modern animal nutrition.

Several plant extracts have shown antimicrobial, anticoccidial, fungicidal or antioxidant properties associated with their lipophylic character (Giannenas *et al.*, 2003; Helander *et al.*, 1998; Juven *et al.*, 1994). Different *in vitro* studies demonstrated a strong inhibition of pathogenic bacteria in the presence of several plant extracts (Dorman and Deans, 2000). In *in vitro* studies by Helander *et al.* (1998), carvacrol, thymol and cinnamaldehyde inhibited the growth of Gram-negative bacteria such as *E. coli* and *Salmonella typhimurium*. According to Ferket *et al.* (2005), antimicrobial ingredients present in phytogenic feed additives may be as effective as AGP in suppressing pathogenic bacteria by either penetrating into the cell or disintegrating the bacterial cell membrane. Lee and Ahn (1998) reported that cinnamaldehyde selectively inhibited *Bacteroides* and *Clostridium perfringens*, which is the causative agent of necrotic enteritis in poultry. Mitsch *et al.* (2004) investigated the efficacies of essential oils in broiler chickens. A blend of thymol, eugenol, curcumin and piperin significantly reduced the concentrations of *Clostridium perfringens* in the intestinal contents and faeces of the birds. The most pronounced effects of supplementation with essential oils were obtained in the first half of the growing period. In *in vitro* studies, Zrustova *et al.* (2005) showed that different essential oils strongly inhibited the growth of *Clostridium perfringens*.

It is generally assumed that phytogenics may stimulate the production of digestive enzymes such as lipase and amylase, thus having a beneficial effect on nutrient utilisation in different categories of animals (Ramakrishna *et al.*, 2003; Williams and Losa, 2001). Finally, they may exert sedative, immuno-stimulating or diuretic effects (Ertas *et al.*, 2005).

## Effects of phytogenics on performance

### EFFECTS OF PHYTOGENICS IN PIGS

In a study with weaned piglets, supplementation of a control diet with a blend of essential oils derived from oregano, anise and citrus peels significantly decreased the number of aerobic and anaerobic bacteria in the ileum and caecum, respectively (Kroismayr *et al.*, 2005). Moreover, there was a significant decrease in the size of Peyer's patches, indicating that the immune system was less active as compared to pigs fed the control diets without phytogenics. In another trial with weaned pigs, supplementation of corn-soybean meal-based diets with a mixture of oregano, cinnamon and Mexican pepper significantly reduced the total number of microorganisms in the ileum and increased the *Lactobacilli/ Enterobacteria* ratio (Manzanilla *et al.*, 2004). Phytogenics may be combined with other NGP in order to obtain synergistic or additive effects with regard to gut health and performance parameters. As reported by Kamel (2000), supplementation of control diets with plant extracts or a blend of plant extracts and organic acids tended to increase feed intake by 2 and 8%, respectively. In the same order, average daily weight gain was significantly improved (8 and 15%). Moreover, feed/gain ratio and incidence of diarrhoea were significantly reduced by both treatments. In piglets weaned at 23 days of age, supplementation of a control diet with a combination of essential oils and FOS increased body weight by up to 12% (Steiner *et al.*, 2006a, Figure 9.1).

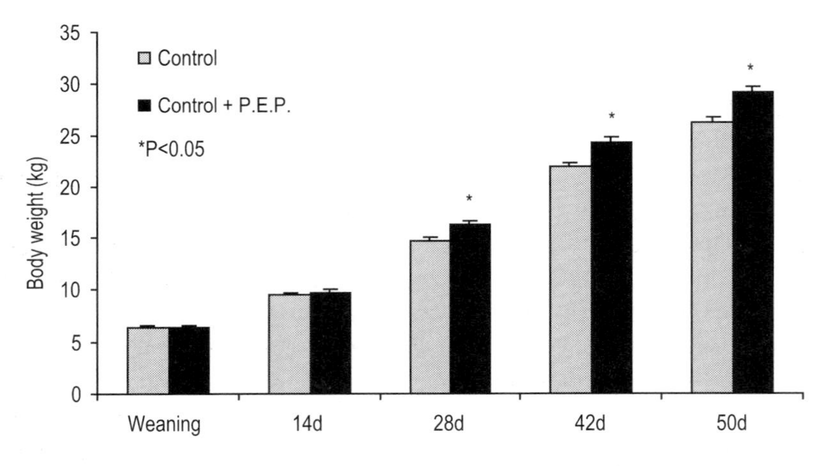


**Figure 9.1.** Effect of a combination of essential oils and FOS (Biomin® P.E.P.) on body weight of piglets (Steiner *et al.*, 2006a)

Furthermore, the phytogenic additive significantly increased feed intake by 8% as compared to the pigs fed the control diet, which has been attributed to its flavouring properties. Finally, there was a tendency towards improved feed efficiency in piglets fed the phytogenic additive. The efficacy of the phytogenic blend used in the trial by Steiner *et al.* (2006a) was also demonstrated in a study with sows (Miller *et al.*, 2003). In this study, phytogenic supplementation of basal diets significantly increased average feed intake during early lactation (5.5 vs. 3.0 kg on day 3 after farrowing) and numerically reduced body weight loss of sows (3.3 vs. 3.5 kg) during the first two weeks after farrowing. Furthermore, piglets from sows fed the phytogenic blend tended to be heavier three weeks after birth (5.8 vs. 5.4 kg).

As can be derived from a large-scale study with 1800 sows (Amrik and Bilkei, 2004), phytogenics have the potential to improve reproduction parameters in sows and piglets. In this study, supplementation of diets for sows with oregano extract significantly lowered mortality, culling rate and significantly increased farrowing rate of the sows, thereby increasing the number of piglets born alive. As reported by Khajarern and Khajarern (2002), supplementation of diets for gestating and lactating sows with essential oils increased body weight at weaning as well as growth performance of piglets, and increased daily feed intake as well as milk yield in sows. In a study with weaned pigs (Velàzquez *et al.*, 2004), a blend of plant extracts originating from oregano, cinnamon and pepper exerted similar effects on performance parameters as compared to an AGP (Carbadox). Both the phytogenic blend and the AGP significantly improved average daily weight gain and feed efficiency in comparison to a negative control treatment.

## EFFECTS OF PHYTOGENICS IN POULTRY

Phytogenic feed additives have also been used in poultry. In studies with broilers, supplementation of corn-soybean meal-based diets with a blend of essential oils significantly increased body weight, feed conversion rate and carcass yield (Alçiçek *et al.*, 2003). In another trial with broilers, administration of oregano essential oils increased (cumulative) feed intake and average live weight by up to 19 and 22%, respectively, and also reduced the counts of *Clostridia perfringens* in the caeca and numbers of coccidia oocysts in the litter (Waldenstedt, 2003). Similar effects of phytogenic supplementation were obtained in birds that were vaccinated against coccidiosis. Ertas *et al.* (2005) conducted a study with broilers fed graded doses (0, 100, 200, 400 ppm) of a blend of essential oils originating from oregano, clove and anise in comparison to an AGP (0.1% Avilamycin). The phytogenic additive significantly improved average daily weight gain and feed conversion ratio and these improvements were even higher in magnitude as compared to the AGP. The highest increments in these parameters (up to 12 and 9%, respectively) were obtained when the phytogenic additive was supplemented at a dosage of 200 ppm, indicating a distinct dose-dependency between performance parameters and the level of supplemental phytogenic additive. Furthermore, the phytogenic mixture stimulated feed intake after three experimental weeks. Similar observations were reported by Sarica *et al.* (2005) with broilers when diets were supplemented with a phytogenic feed additive based on thyme. In female turkeys, Bampidis *et al.* (2005) obtained a linear increase in feed efficiency when the birds were fed increasing levels of dried oregano leaves.

## EFFECTS OF PHYTOGENICS IN RUMINANTS

Due to the antimicrobial properties of different herbs and spices, the use of phytogenics in ruminant nutrition may allow for manipulation of the ruminal microbial fermentation in order to improve the efficiency of nutrient utilisation. However, suitable results from *in vivo* studies are still missing. In *in vitro* experiments, Busquet *et al.* (2006) investigated the impact of different plant extracts and secondary plant metabolites on rumen fermentation parameters and reported that most of the substances under investigation decreased the concentrations of volatile fatty acids in the rumen fluid. Moreover, some of the substances (e.g. anethol, anise oil or tea tree oil) decreased the acetate/propionate ratio, indicating a non-beneficial impact on ruminal fermentation. Similar observations were reported by Cardozo *et al.* (2005). These authors, however, indicated that the effect of phytogenic agents on ruminal fermentation depends on the actual pH conditions in the rumen. Some compounds, including oregano, capsicum or cinnamaldehyde, significantly reduced the acetate/propionate ratio and lowered the production of volatile fatty acids and ammonia, thus indicating an improved utilisation of nutrients and dietary energy. In a study with heifers by Hristov *et al.* (1999), intraruminal application of powdered Yucca extracts significantly decreased ruminal ammonia concentration and number of protozoa, coinciding with increased concentrations of propionate in comparison to a control treatment. However, nutrient digestibilities and ruminal protein synthesis were not affected by infusion of Yucca extracts.

# 10

# IMMUNE STIMULANTS

## Characterisation

The gut mucosa is the largest contact surface between the host and the environment, thus representing a potential entrance gate for pathogens to invade the body. Healthy animals are protected against pathogenic invasion by a highly efficient immune system. A large number of immune cells is directly or indirectly associated with the gut, modulating immune responses towards induction or tolerance (Goddeeris, 2005). According to Bar-Shira *et al.* (2003), the development of the gut-associated immune system is concomitant with the structural and functional maturing of the gut. Hence, in young animals, the acquired immune system starts to develop rather slowly after birth, thus making them susceptible to gastrointestinal infections. Particularly in times of stress, the immune system functions rather weakly, due to stress-induced immune suppression (Allen *et al.*, 2001). It has been known for many years that the gut-associated immune system can be modulated by nutritional means (Adams, 2001). There are many studies evaluating strategies to support the immune system by inclusion of suitable immune-stimulating or -modulating agents in animal feeds. Among others, cell wall fragments originating from yeasts or specific bacteria have been shown to assist in stimulating mechanisms of the innate immunity, for example by activating macrophages, lymphocytes or natural killer cells (Goddeeris, 2005). Also, specific probiotic bacteria (e.g. *Lactobacilli*) and phytogenic compounds originating from herbs and spices as well as aquatic plants may exert beneficial effects on the host immune system.

## Effects of immune stimulants on immune status

A well-balanced gut microflora is essential for adequate and rapid development of the gut-associated immune system. The positive impact of probiotic supplementation on immune status has been investigated in aquatic species. Panigrahi *et al.* (2005) obtained a significant increase in phagocytic activity of kidney leucocytes and plasma immune globulin concentrations in rainbow trout when control diets were supplemented with *Lactobacillus rhamnosus*. Positive effects of microorganisms on the activity of different immune cells may be related particularly to specific cell wall components such as peptidoglucans, ß-1,3-1,6-glucans or mannan oligosaccharides (MOS).

According to Samegai (2000), peptidoglucans derived from *Bifidobacterium thermophilum* may induce enhanced cellular immune response in piglets. Earlier, Sasaki *et al.* (1987) fed peptidoglucan originating from cell walls of *Bifidobacterium thermophilum* to weaned piglets and observed a significant increase in immune globulin A-bearing cells of the lamina propria, which is a connective tissue of the ileum and jejunum containing a large number of different immune cells. Moreover, De Ambrosini *et al.* (1998) demonstrated that cell walls and, in particular, peptidoglucans originating from *Lactobacillus casei* bacteria markedly stimulated phagocytic cells in mice. In the same study, administration of *Lactobacillus acidophilus* increased the levels of immune globulin A antibodies in the intestinal fluid. Increased phagocytosis was also observed in yellowtail (*Seriola quinqueradiata*) by oral administration of peptidoglucans originating from *Bifidobacterium thermophilum*. Several studies, however, failed to show a significant effect of microbial cell wall components on immune and performance parameters in livestock.

The impact of MOS, derived from cell walls of *Saccharomyces cerevisiae*, on immune status and growth performance has been investigated in a large number of trials. The reported efficacy of MOS supplementation has been attributed to the potential of MOS to act as a ligand offering a competitive binding site for potentially harmful bacteria that have mannose-specific Type-1 fimbriae receptors on their surface. It is assumed that the binding of pathogenic bacteria to MOS prevents these bacteria from attaching to the gut wall. Under *in vitro* conditions, Gram-negative bacteria such as *Salmonella* or *E. coli* were indeed agglutinated by MOS (Ofek *et al.*, 1977). Similarly, adherence of *Salmonella typhimurium* to the small intestinal tissues of chicken *in vitro* was inhibited when mannose was present in the incubation medium (Oyofo *et al.*, 1989). However, as can be derived from published literature, the impact of MOS under *in vivo* conditions on performance, pathogenic counts or immune cells is rather inconsistent (Burkey *et al.*, 2004; LeMieux *et al.*, 2003; White *et al.*, 2002). Positive effects on these parameters have been obtained, for example, when animals were artificially challenged with pathogenic bacteria (e.g. Burkey *et al.*, 2004; Davis *et al.*, 2004). In contrast, no effect of dietary supplementation of MOS on these parameters was reported by Ko *et al.* (2000) and White *et al.* (2002).

As reviewed by Allen *et al.* (2001), specific sea plants such as brown seaweed (*Ascophyllum nodosum*) may exert beneficial effects on the immune status and growth performance of livestock, presumably due to distinct antioxidative and antimicrobial properties of these organisms. In studies of Behrends *et al.* (2000), feeding of brown seaweed to Angus steers reduced the levels *E. coli* in faecal and hide samples. Furthermore, seaweed has shown antitumour effects in rodents (Teas *et al.*, 1984; Yamamoto *et al.*, 1977). Turner *et al.* (2002) observed a quadratic effect on average daily weight gain and final body weight, as well as a linear effect on average daily feed intake and gain/feed ratio when pigs challenged with *Salmonella typhimurium* were fed increasing amounts of brown seaweed. The impact of sea plants and algae extracts was also investigated in aquatic species. Especially during the process of rearing, which is extremely stressful for fish larvae, several immune-modulating agents may be beneficial to strengthen the immune response. Oral administration of sulfated D-galactans extracted from red marine algae (*Botryocladia occidentalis*) significantly increased growth rates in tilapia larvae (Farias *et al.*, 2004). Similarly, supplementation of basal feed with peptidoglucans improved growth rates and feed conversion in black tiger shrimp (Boonyaratpalin *et al.*, 1995).

It can be derived from the above-mentioned studies that cell wall constituents originating from bacteria or yeasts may have a beneficial impact on the immune status of livestock when administered in the feed. Under practical conditions, however, positive effects may probably only be expected when animals are subjected to stress, i.e. if the hygienic standards are poor and the pathogenic pressure is high.

Apart from the substances discussed in the previous chapters, there are several other agents that may induce or maintain immune response of the host animal, including vitamins A and D, poly-unsaturated fatty acids (omega-3 and -6 fatty acids) and L-carnitine (Goddeeris, 2005). In conclusion, there is increasing knowledge of the complexity of interactions between nutrition and immunity which may provide suitable strategies to modulate immune system by nutritional means. However, the modes of action of different immune-modulating agents require further investigation.

## Effects of immune stimulants on performance

### EFFECTS OF IMMUNE STIMULANTS IN PIGS

A report by Dritz *et al.* (1995), in which the impact of ß-glucans originating from yeast (*Saccharomyces cerevisiae*) cell walls on immune status and growth performance of piglets was investigated, revealed inconsistent results. In a first experiment, when piglets were fed diets based on milk protein, supplementation of ß-glucan (0.1% of diet) significantly decreased average daily gain and average daily feed intake by 17 and 21%, respectively. In another experiment, however, when pigs were fed diets based on soy protein, addition of ß-glucans at a lower dosage (0.025% of diet) significantly improved these parameters by up to 21 and 23%, respectively. These results indicate a close relationship between dietary dosage of immune-stimulating microbial cell walls and performance parameters. According to Goddeeris (2005), induction and maintenance of immune mechanisms is always at a cost of metabolic energy which is, in turn, lost for productive purposes. Thus, excessive stimulation of immune cells may result in reduced growth performance. Since ß-glucans may elicit specific immune reactions by stimulating the secretion of interleukins, it seems likely that excessive activation of immune response limited growth performance in the first experiment by Dritz *et al.* (1995). However, responses to oral administration of ß-glucans were marginal in several studies. Burkey *et al.* (2004) conducted a trial with pigs that were orally challenged with *Salmonella enterica* to evaluate the potential of MOS on immune response and growth performance. Addition of MOS to control diets tended to increase serum concentration of haptoglobin, indicating a slight stimulation of the acute-phase immune response. However, there was no marked influence of MOS supplementation on growth performance or concentrations of pro-inflammatory cytokines (such as inteleukin-6), which play an essential role in the communication of different immune cells. In a study by Davis *et al.* (2004) with 19-day-old piglets, dietary supplementation of phosphorylated MOS significantly increased the concentration of lymphocytes as well as average daily gain and gain/feed ratio in the first two weeks of the experiments. Addition of MOS to diets for sows had no effect on sow body weight loss during lactation and number of piglets born alive, but increased litter weight at birth (O'Quinn *et al.*, 2001).

## EFFECTS OF IMMUNE STIMULANTS IN POULTRY

In trials with broilers, Fleischer *et al.* (2000) observed negligible effects of dietary supplementation with ß-glucans on the activity of different immune cells. Den Hartog *et al.* (2005) carried out a study with broilers that were either challenged or not challenged with digesta homogenate taken from birds suffering from malabsorption syndrome, which is a widespread disease causing growth depression, pale skin colouring and inadequate skeletal development. The authors reported that supplementation of the basal diets with yeast ß-glucans improved growth performance only in the infected birds and during the first three weeks of the experiment.

## EFFECTS OF IMMUNE STIMULANTS IN AQUACULTURE

Infectious diseases are a potential threat in modern, intensive aquaculture systems. A comprehensive review has been published earlier by Sakai (1999), indicating that several substances, including synthetic chemicals, bacterial derivates, animal and plant extracts, vitamins, hormones or cytokines, have the potential to protect aquatic species from severe infections. Li and Gatlin (2005) investigated the impact of immune stimulating yeast on growth performance in hybrid striped bass (*Morone chrysops* × *M. saxatilis*). Addition of 2% brewer's yeast to the control diet significantly increased weight gain after 4 and 2 weeks by 8 and 15%, respectively, and also improved feed efficiency significantly by 5% after 16 weeks. Wang and Wang (1997) investigated the efficacies of different polysaccharides in tilapia (*Tilapia aureus*) and grass carp (*Ctenopharyngodon idellus*) after infection with bacterial pathogens (either *Aeromonas hydrophila* or *Edwardsiella tarda*). In this study, ß-glucans of different microbial origins and with different binding structures (fungal ß-1,3-1,4-glucan, fungal ß-1,4-1,3-1,6-glucan, fungal ß-1,3-1,6-glucan or yeast ß-glucan-protein-lipid complex) significantly increased survival rates of the fish, indicating that these polysaccharides may promote resistance against the above-mentioned bacterial infections by inducing a non-specific immune response.

## EFFECTS OF IMMUNE STIMULANTS IN RUMINANTS

The efficacy of MOS was also investigated in ruminants. In a study with 24 heifer calves, supplementation of a control diet with MOS improved average daily weight gain by 3.4% (Shimkus, 2004). Moreover, MOS increased the number of infusoria and bacteria in the rumen fluid by 36 and 69%, respectively, indicating a potential of MOS to affect the pattern of ruminal nutrient degradation.

# 11

# CONCLUSION

Gut health is mainly based on a well-balanced microflora which protects the host from pathogenic invasion and, thus, has a beneficial impact on the overall health status of the animal. In the past, AGP were routinely included in diets for food-producing animals in order to prevent gastrointestinal disorders and maintain high performance levels. Since the ban of AGP in the European Union has been implemented in January 2006, alternative concepts have been developed in order to maintain gut health. A suitable combination of several natural feed additives has the potential to compete against unfavourable dietary factors such as pathogenic contamination, change in dietary composition or stress. These additives, referred to as Natural Growth Promoters (NGP), include organic acids, probiotics, prebiotics, feed enzymes, phytogenic feed additives and immune stimulants. The main mode of action of organic acids, probiotics, prebiotics and phytogenics is that they modify and stabilise the gut microflora by providing comfortable conditions for beneficial microorganisms, while inhibiting the growth of pathogenic species.

The mechanisms involved are by far from completely understood and, therefore, require further intensive research. The use of modern molecular techniques for determination of quantitative and qualitative aspects of microbial communities provides a promising tool for further discovery of the mechanisms involved in the action of NGP. Previously, decrease in gut pH, modification of gut physiology and morphology, alterations of nutrient availability and stimulation of the immune system have been identified as important mechanisms determining the efficacy of NGP under *in vivo* conditions. In most cases, the effects of supplementation of NGP seem to be dose-dependent. There are usually great differences in efficacy regarding the mode of action as well as effects on microflora, digestibility, performance and immunity, between different products within the same category that are available on the market. These differences are mainly due to a large variation in (chemical) composition of these products. Moreover, the *in vivo* efficacy of NGP depends on several additional factors including, among others, age of animals, dietary composition, dosage of NGP, buffering capacity and hygienic conditions. Further research should examine the factors determining the efficacies of different NGP alone and in combinations under *in vivo* conditions to optimise their use in different categories of animals.

Apart from pigs, poultry and ruminants, the implication of different feeding strategies based on NGP will play a vital role in replacing AGP also in a wide range of aquaculture species, thus following the requests of officials and, more importantly, the consumers for safe and high-quality foods.

In conclusion, NGP are a suitable alternative to AGP, and may assist to increase energy and nutrient availability, improve growth performance and reduce the incidence of gastrointestinal disorders. However, their efficacy determined in *in vitro* experiments has to be confirmed under practical conditions *in vivo*. Finally, a well-balanced combination of different products may optimise the benefits of NGP. For example, combination of probiotics, prebiotics and immune-stimulating agents may successfully protect the animal from pathogenic invasion, thereby promoting overall intestinal health and growth performance. Under certain environmental conditions (e.g. poor hygienic conditions and high pathogenic pressure), it might be appropriate to use a combination of acidifiers and phytogenics, whereas in other cases, probiotics in combination with prebiotics may be the alternative of choice. Therefore, optimal strategies should be carefully adapted to individual feeding, management and hygienic conditions in order to optimise the efficiency of NGP in antibiotic-free feeding systems.

# 12

# SUMMARY

The ban of antibiotic growth promoters as feed additives in the European Union has forced researchers and the feed industry to develop suitable alternative strategies in order to maintain animal health and growth performance at a high level. Several Natural Growth Promoters (NGP), including organic acids, probiotics, prebiotics, feed enzymes, phytogenics and immune-stimulating substances, have been evaluated for their potential to support animal health and to improve weight gain and feed efficiency in different categories of food-producing animals.

(1) Diet acidification is widely used to avoid microbial degradation of feeds during storage. Organic acids may improve gastric proteolysis, inhibit the proliferation of pathogenic bacteria and, finally, increase growth performance. (2) Probiotics are live microorganisms that modify the gut microflora in a beneficial way, thereby inhibiting the attachment and proliferation of pathogenic bacteria. Thus, they have the potential to prevent intestinal disorders and diseases, especially in young animals and after antibiotic treatment. (3) Prebiotics represent a specific substrate for intestinal microorganisms. These fermentable carbohydrates are not digestible by endogenous enzymes but they selectively stimulate the growth of beneficial bacteria, thus promoting and maintaining a healthy gut microflora. Probiotics and prebiotics are often combined as synbiotics in order to obtain synergistic or additive effects of these categories of NGP. (4) Feed enzymes are supplemented to the feed to increase digestibility of non-starch polysaccharides or phytate phosphorus, or to support the activity of endogenous proteolytic or amylolytic enzymes in young animals. Moreover, they may affect the gut microflora and intestinal morphology. (5) Due to aromatic properties, phytogenics originating from herbs, spices and other plants may stimulate feed intake in various animal species. Additionally, they have strong antimicrobial effects and, therefore, reduce the competition for nutrients between the host and its gut microflora. (6) Finally, immune-modulating agents may be used as a valuable tool to prevent infectious diseases by inducing immune response, particularly in times of stress, when the host immune system is constricted.

The *in vivo* efficacy of NGP is affected by several factors, such as animal species, dietary composition, dosage of NGP, buffering capacity and hygienic conditions. In general, the benefits of NGP are highest in young animals, due to their limited digestive capacity which is characterised by an unstable gut microflora, inadequate endogenous enzyme secretion and immature immune system. Future strategies should aim to combine different NGP in a skilful way in order to optimise their benefits in antibiotic-free feeding systems, taking into account the above-mentioned factors.

# 13

# REFERENCES

Abu-Tarboush, H.M., Al-Saiady, M.Y. and Keir El-Din, A.H. (1996) Evaluation of diet containing *lactobacilli* on performance, fecal coliform, and *lactobacilli* of young dairy calves. *Animal Feed Science and Technology* **57**: 39–49.

Adams, C.A. (2001) Health promoting additives (nutricines). *1st World Feed Conference: Advances in Nutritional Technology,* Utrecht, The Netherlands, pp. 207–227.

Alçiçek, A., Bozkurt, M. and Çabuk, M. (2003) The effect of an essential oil combination derived from selected herbs growing wild in Turkey on broiler performance. *South African Journal of Animal Science* **33**: 89–94.

Alexopoulos, C., Georgoulakis, I.E., Tzivara, A., Kritas, S.K., Siochu, A. and Kyriakis, S.C. (2004) Field evaluation of the efficacy of a probiotic containing *Bacillus licheniformis* and *Bacillus subtilis* spores, on the health status and performance of sows and their litters. *Journal of Animal Physiology and Animal Nutrition* **88**: 281–292.

Allen, V.G., Pond, K.R., Saker, K.E., Fontenot, J.P., Bagley, C.P., Ivy, R.L., Evans, R.R., Schmidt, R.E., Fike, J.H., Zhang, X., Ayad, J.Y., Brown, C.P., Miller, M.F., Montgomery, J.L., Mahan, J., Wester, D.B. and Melton C. (2001) Tasco: Influence of a brown seaweed on antioxidants in forages and livestock – A review. *Journal of Animal Science* **79** (E. Suppl.): E21–E31.

Amrik, B. and, Bilkei, G. (2004) Influence of farm application of oregano on performance of sows. *Canadian Veterinary Journal-Revue Veterinaire Canadienne* **45**: 674–677.

Åsgård, T., and Austreng, E. (1981) Fish silage for salmonids: a cheap way of utilizing waste as feed. *Feedstuffs* **53**: 22–24.

Bach Knudsen, K.E. (1997) Carbohydrate and lignin contents of plant materials used in animal feeding. *Animal Feed Science and Technology* **67**: 319–338.

Bailey, R.W. (1973) Structural carbohydrates. In: *Chemistry and Biochemistry of Herbage*: No. 1. Edited by Butler, G.W., Bailey, R.W. pp. 157–211. Academic Press, New York.

Bampidis, V.A., Christodoulou, V., Florou-Paneri, P., Christaki, E., Chatzopoulou, P.S., Tsiligianni, T. and Spais, A.B. (2005) Effect of dietary dried oregano leaves on growth performance, carcase characteristics and serum cholesterol of female early maturing turkeys. *British Poultry Science* **46**: 595–601.

Bar-Shira, E., Sklan, D. and Friedman, A. (2003) Establishment of immune competence in the avian GALT during the immediate post-hatch period. *Developmental and Comparative Immunology* **27**: 147–157.

Bauer, E., Williams, B.A., Verstegen, M.W.A. and Mosenthin, R. (2006) Fermentable carbohydrates: potential dietary modulators of intestinal physiology, microbiology and immunity in pigs. In: *Biology of Growing Animals Series*: Vol. 4. Biology of Nutrition in Growing Animals. Edited by Mosenthin, R., Zentek, J., Zebrowska, T. pp. 33–63. Elsevier Limited, Edinburgh, United Kingdom.

Beauchemin, K.A., Colombatto, D. and Morgavi, D.P. (2004) A rationale for the development of feed enzyme products for ruminants. *Canadian Journal of Animal Science* **84**: 23–36.

Bedford, M.R. and Apajalahti, J. (2001) Implications of diet and enzyme supplementation on the microflora of the intestinal tract. In: *Advances in Nutritional Technology.* Proceeding of the 1st World Feed Conference. Edited by van der Poel, A.F.B., Vahl, J.L., Kwakkel, P.R. pp. 197–206. Utrecht, The Netherlands.

Behrends, L.L., Blanton, J.R. Jr., Miller, M.F., Pond, K.R. and Allen, V.G. (2000) Tasco supplementation in feedlot cattle: Effects on pathogen loads. *Journal of Animal Science* **78** (Suppl. 1): 106.

Berchieri, A., Sterzo, E., Paiva, J., Lückstädt, C. and Beltran, R. (2006). The use of a defined probiotic product (Biomin® PoultryStar) and organic acids to control *Salmonella enteritidis* in broiler chickens. *International Poultry Scientific Forum*, Atlanta, USA, 23 – 24 January 2006, p. 30.

Biagi, G., Piva, A., Hill, T., Schneider, D.K. and Crenshaw, T.D. (2003) Low buffering capacity diets with added organic acids as a substitute for antibiotics in diets for weaned pigs. *9th International Seminar on Digestive Physiology in the Pig*: Vol. 2., pp. 217–219.

Blank, R., Mosenthin, R., Sauer, W.C. and Huang, S. (1999) Effect of fumaric acid and dietary buffering capacity on ileal and fecal amino acid digestibilities in early-weaned pigs. *Journal of Animal Science* **77**: 2974–2984.

Bolduan, G., Jung, H., Schneider, R., Block, J. and Klenke, B. (1988) Effect of propionic and formic acid in piglets. *Journal of Animal Physiology and Animal Nutrition* **59**: 72–78.

Boling, S.D., Webel, D.M., Mavromichalis, I., Parsons, C.M. and Baker, D.H. (2000) The effects of citric acid on phytate-phosphorus utilization in young chicks and pigs. *Journal of Animal Science* **78**: 682–689.

Boonyaratpalin, S., Boonyaratpalin, M., Supamattaya, K. and Yoride, Y. (1995) Effects of peptidoglucan (PG) on growth, survival, immune responses, and tolerance to stress in black tiger shrimp, *Penaeus monodon*. In: *Diseases in Asian Aquaculture*. Vol. 11. Fish Health Section, Asian Fisheries Society. Edited by Shariff, M., Subasighe, R.P., Arthur, J.R. pp. 469–477. Manila, Philippines.

Burkey, T.E., Dritz, S.S., Nietfeld, J.C., Johnson, B.J. and Minton, J.E. (2004) Effect of dietary mannanoligosaccharide and sodium chlorate on the growth performance, acute-phase response, and bacterial shedding of weaned pigs challenged with *Salmonella enterica* serotype Typhimurium. *Journal of Animal Science* **82**: 397–404.

Busquet, M., Calsamiglia, S., Ferret, A. and Kamel, C. (2006) Plant extracts affect *in vitro* rumen microbial fermentation. *Journal of Dairy Science* **89**: 761–771.

Canh, T.T., Sutton, A.L., Aarnink, A.J., Verstegen, M.W.A., Schrama, J.W. and Bakker, G.C. (1998) Dietary carbohydrates alter the fecal composition and pH and the ammonia emission from slurry of growing pigs. *Journal of Animal Science* **76**: 1887–1895.

Canibe, N., Steien, S.H., Øverland, M. and Jensen, B.B. (2001) Effect of K-diformate to starter piglet diets on digesta and faecal microbial profile, and on stomach alterations. *Journal of Animal Science* **79**: 2123–2133.

Cardozo, P.W., Calsamiglia, S., Ferret, A. and Kamel, C. (2005) Screening for the effects of natural plant extracts at different pH on in vitro rumen microbial fermentation of a high-concentrate diet for beef cattle. *Journal of Animal Science* **83**: 2572–2579.

Cebra, J.J. (1999) Influences of microbiota on intestinal immune system development. *American Journal of Clinical Nutrition* **69**:1046S–1051S.

Chambers, J.R., Spencer, J.L. and Modler, H.W. (1997) The influence of complex carbohydrates on *Salmonella typhimurium* colonization, pH, and density of broiler ceca. *Poultry Science* **76**: 445–451.

Chen, Y.C., Nakthong, C. and Chen, T.C. (2005) Improvement of laying hen performance by dietary prebiotic chicory oligofructose and inulin. *International Journal of Poultry Science* **4**: 103–108.

Choct, M. (2004) Enzymes for the feed industry: Past, present and future. *XXII World's Poultry Congress*, Istanbul, Turkey, pp. 1–9.

Choct, M. (1997) Feed Non-Starch Polysaccharides: Chemical structures and nutritional significance. *Feed Milling International*, June issue, 13–26.

Choct, M., Kocher, A., Waters, D.L.E., Pettersson, D. and Ross, G. (2004) A comparison of three xylanases on the nutritive value of two wheats for broiler chickens. *British Journal of Nutrition* **92**: 53–61.

Collier, C.T., Smiricky-Tjardes, M.R., Albin, D.M., Wubben, J.E., Gabert, V.M.,

Deplancke, B., Bane, D., Anderson, D.B. and Gaskins, H.R. (2003) Molecular ecological analysis of porcine ileal microbiota responses to antimicrobial growth promoters. *Journal of Animal Science* **81**: 3035–3045.

Conway, P.L. (1996) Development of the intestinal microbiota. Gastrointestinal microbes and host interactions. In: *Gastrointestinal Microbiology*: Vol. 2. Edited by Mackie, R.I., White, B.A., Isaacson, R.E. pp. 3–39. Chapman and Hall, London.

Conway, P.L. (1994) Function and regulation of the gastrointestinal microbiota of the pig. In: *Proceedings of the VI*[th] *International Symposium on Digestive Physiology in Pigs.* EAAP Publication no. 80. Edited by Souffrant, W.B., Hagemeister, H. pp. 231–240.

Cranwell, P.D. (1985) The development of acid and pepsin (EC 3.4.23.1) secretory capacity in the pig: the effects of age and weaning. 1. Studies in anaesthetized pigs. *British Journal of Nutrition* **54**: 305–320.

Cromwell, G.L. (1991) Anti-microbial agents. In: *Swine Nutrition.* Edited by Miller, E.R., Ullrey, D.E., Lewis, A.J. pp. 297–314. Butterworth-Heinemann, Boston, USA.

Dänicke, S. (1999) Zum Einfluß von Nicht-Stärke-Polysacchariden (NSP) und NSP-spaltenden Enzymen auf die Passagezeit der Ingesta sowie den Energie- und Proteinumsatz von wachsenden Schweinen und Broilern. (On the influence of non-starch-polysaccharides (NSP) and NSP-hydrolyzing enzymes on transit time of ingesta and energy- and protein metabolism of growing pigs and broilers). *Übersichten zur Tierernährung* **27**: 221–273.

Daniel, T.C., Sharpley, A.N. and Lemunyon, J.L. (1998) Agricultural phosphorus and eutrophication: A symposium overview. *Journal of Environmental Quality* **2**: 251–257.

Danmap (2002) Danmap 2002 - Use of antimicrobial agents and occurrence of antimicrobial resistance in bacteria from food animals, foods and humans in Denmark. Danish Institute for Food and Veterinary Research, Copenhagen, Denmark.

Davis, M.E., Maxwell, C.V., Erf, G.F., Brown, D.C. and Wistuba, T.J. (2004) Dietary supplementation with phosphorylated mannans improves growth response and modulates immune function of weanling pigs. *Journal of Animal Science* **82**: 1882–1891.

De Ambrosini, V.M., Gonzales, S., Perdigon, G., De Ruiz Holgado, A.P. and Oliver, G. (1998) Immunostimulating activity of cell walls from lactic acid bacteria and related species. *Food and Agricultural Immunology* **10**: 183–191.

Decamp, O., van de Braak, K. and Moriarty, D.J.W (2005) Probiotics for shrimp larviculture - Review of field data from Asia and Latin America. *Larvi 2005. 4th Fish & Shellfish Larviculture Symposium*, September 5-8, 2005, Gent, Belgium. Abstract CD-Rom.

Den Hartog, L.A., Gutierrez del Alamo Oms, A., Doorenbos, J. and Flores Minambres, A. (2005) The effect of natural alternatives for anti-microbial growth promoters in broiler diets. *15th European Symposium on Poultry Nutrition*, 24–29 September 2005, Balatonfüred, Hungary, pp. 212–220.

De Wet, L. (2005) Can organic acid effectively replace antibiotic growth promotants in diets for rainbow trout *Oncorhynchus mykiss* raised under sub-optimal water temperatures? *WAS Conference*, May 9–13, 2005, Bali, Indonesia. Abstract CD-Rom.

Diebold, G. (2005) Supplementation of xylanase and phospholipase to wheat-based diets for weaner pigs. Doctoral thesis, University of Hohenheim, Stuttgart, Germany.

Diebold, G., Mosenthin, R., Piepho, H.-P. and Sauer, W.C. (2004) Effect of supplementation of xylanase and phospholipase to a wheat-based diet for weanling pigs and nutrient digestibility and concentrations of microbial metabolites in ileal digesta and feces. *Journal of Animal Science* **82**: 2647–2656.

Dhiman, T.R., Zaman, M.S., Gimenez, R.R., Walters, J.L. and Treacher, R. (2002) Performance of dairy cows fed forage treated with fibrolytic enzymes prior to feeding. *Animal Feed Science and Technology* **101**: 115–125.

Dilger, R.N., Onyango, E.M., Sands, J.S. and Adeola, O. (2004) Evaluation of microbial phytase in broiler diets. *Poultry Science* **83**: 962–970.

Domig, K.J. (2005) Antibiotikaresistenz und der Einsatz von Antibiotika in der Tierernährung. *4. BOKU-Symposium Tierernährung: Tierernährung ohne Antibiotische Leistungsförderer*. Vienna, Austria, pp.1–8.

Dorman, H.J.D. and Deans, S.G. (2000) Antimicrobial agents from plants: antibacterial activity of plant volatile oils. *Journal of Applied Microbiology* **88**: 308–316.

Dritz, S.S., Shi, J., Kielian, T.L., Goodband, R.D., Nelssen, J.L., Tokach, M.D., Chengappa, M.M., Smith, J.E. and Blecha, F. (1995) Influence of dietary beta-glucan on growth performance, nonspecific immunity, and resistance to *Streptococcus suis* infection in weanling pigs. *Journal of Animal Science* **73**: 3341–3350.

Dusel, G., Schulze, H., Kluge, H., Simon, O. and Jeroch, H. (1997) The effect of wheat quality measured by extract viscosity and dietary addition of feed enzymes on performance of young pigs. *Journal of Animal Science* **75** (Suppl. 1): 200.

Eckel, B., Kirchgessner, M. and Roth, F.X. (1992) Zum Einfluß von Ameisensäure auf tägliche Zunahmen, Futteraufnahme, Futterverwertung und

Verdaulichkeit. 1. Mitteilung: Untersuchungen zur nutritiven Wirksamkeit von organischen Säuren in der Ferkelaufzucht. *Journal of Animal Physiology and Animal Nutrition* **67**: 93–100.

Eeckhout, W. and De Paepe, M. (1994) Total phosphorus, phytate-phosphorus and phytase activity in plant feedstuffs. *Animal Feed Science and Technology* **47**: 19–29.

Eidelsburger, U. (1997) Organische Säuren und was sie in der Schweinefütterung bewirken. Optimierung der Futterqualität ist nur ein Teilaspekt. *Schweinewelt*, Jahrgang 22, 1/97, 18–21.

Engberg, R.M., Hedemann, M.S., Leser, T.D. and Jensen, B.B. (2000) Effect of zinc bacitracin and salinomycin on intestinal microflora and performance of broilers. *Poultry Science* **79**: 1311–1319.

Engelen, A.J., van der Heeft, F.C., Randsdorp, P.H.G. and Smit, E.L.C. (1994) Simple and rapid determination of phytase activity. *Journal of AOAC International* **77**: 760–764.

Englyst, H.N., Bingham, S.A., Runswick, S.A., Collins, E. and Cummings, J.H. (1989) Dietary fiber (non-starch-polysaccharides) in cereal products. *Journal of Human Nutrition and Dietetics* **2**: 253–271.

Erasmus, L.J., Botha, P.M. and Kistner, A. (1992) Effect of yeast culture supplement on production, rumen fermentation, and duodenal nitrogen flow in dairy cows. *Journal of Dairy Science* **75**: 3056–3065.

Erdman, J.W. Jr. (1979) Oilseed phytates: nutritional implications. *Journal of the American Oil Chemists Society* **56**: 736–741.

Ertas, O.N., Güler, T., Çiftçi, M., DalkIlIç, B. and Simsek, Ü.G. (2005) The effect of an essential oil mix derived from oregano, clove and anise on broiler performance. *International Journal of Poultry Science* **4**: 879–884.

Ewing, W.N. and Cole, D.J.A. (1994) The gastrointestinal tract. In: *The living gut: An Introduction to Micro-Organisms in Nutrition.* Edited by Ewing, W.N., Cole, D.J.A. pp. 10–28. Nottingham University Press, UK: Context Publication.

Farias, W.R.L., Rebouças, H.J., Torres, V.M., Rodrigues, J.A.G., Pontes, G.D.A., da Silva, F.H.O. and Sampaio, A.H. (2004) Enhancement of growth in tilapia larvae (*Oreochromis niloticus*) by sulfated D-galactans extracted from the red marine alga *Botryocladia occidentalis*. *Revista Ciência Agronômica* **35**: 189–195.

Farnworth, E.R., Modler, H.W., Jones, J.D., Cave, N., Yamazaki, H. and Rao, A.V. (1992) Feeding Jerusalem artichoke flour rich in fructooligosaccharides to weanling pigs. *Canadian Journal of Animal Science* **72**: 977–980.

Feng, P., Hunt, C.W., Pritchard, G.T. and Julien, W.E. (1996) Effect of enzyme preparations on in situ and in vitro degradation and in vivo digestive characteristics of mature cool-season grass forage in beef steers. *Journal of Animal Science* **74**: 1349–1357.

Ferket, P.R., Santos, A.A. Jr. and Oviedo-Rondón, E.O. (2005) Dietary factors that affect gut health and pathogen colonization. *32nd Annual Carolina Poultry Nutrition Conference,* Research Triangle Park, North Carolina, USA, pp. 1–22.

Fleischer, L.G., Gerber, G., Liezenga, R.W., Lippert, E., Scholl, M.A. and Westphal, G. (2000) Blood cells and plasma proteins of chickens fed a diet supplemented with (1→3),(1→6)-beta-D-glucan and enrofloxacin. *Archiv für Tierernährung* **53**: 59–73.

Foegeding, P.M. and Busta, F.F. (1991) Chemical food preservatives. In: *Disinfection, Sterilization and Preservation*. Edited by Block, S.S. pp. 802–832 Philadelphia, PA: Lea and Febiger.

Foote, K. (2003) The battle of the bugs and other alternatives to antibiotics in pork production. *MB Swine Seminar 2003*: Vol. 17. pp. 1–17.

Francesch, M., Broz, J. and Brufau, J. (2005) Effects of an experimental phytase on performance, egg quality, tibia ash content and phosphorus bioavailability in laying hens fed on maize- or barley-based diets. *British Poultry Science* **46**: 340–348.

Freitag, M., Hensche, H.U., Schulte-Sienbeck, H. and Reichelt, B. (1999) Biologische Effekte konventioneller und alternativer Leistungsförderer. *Kraftfutter/Feed Magazine* **2**: 49–57.

Fuchs E., Binder, E.M., Heidler, D. and Krska, R. (2002) Structural characterisation of metabolites after the microbial degradation of A-trichothecenes by the bacterial strain BBSH 797. Food Additives and Contaminants **19**: 379–386.

Fuller, R. (1989) Probiotics in man and animals. *Journal of Applied Bacteriology* **66**: 365–378.

Gabert, V.M. and Sauer, W.C. (1995) The effect of fumaric acid and sodium fumarate supplementation to diets for weanling pigs on amino acid digestibility and volatile fatty acid concentrations in ileal digesta. *Animal Feed Science and Technology* **53**: 43–254.

Gauthier, R. (2005) Organic acids and essential oils, a realistic alternative to antibiotic growth promoters. *I Forum Internacional de Avicultura*, August 17–19 2005, Foz do Iguaçu, PR, Brasil, pp. 148 – 157.

Gebert, S., Bee, G., Pfirter, H.P. and Wenk, C. (1999) Phytase and vitamin E in the feed of growing pigs: 1. Influence on growth, mineral digestibility and fatty acids in digesta. *Journal of Animal Physiology and Animal Nutrition* **81**: 9–19.

Giannenas, I., Florou-Paneri, P., Papazahariadou, M., Christaki, E., Botsoglou,

N.A. and Spais, A.B. (2003) Effect of dietary supplementation with oregano essential oil on performance of broilers after experimental infection with *Eimeria tenella. Archives of Animal Nutrition* **57**: 99–106.

Gibson, G.R. and Roberfroid, M.B. (1995) Dietary modulation of the human colonic microbiota: introducing the concept of prebiotics. *Journal of Nutrition* **125**: 1401–1412.

Giesting, D.W. (1986) Utilization of soy protein by the young pig. Doctoral thesis, University of Illinois, Urbana, Champaign, USA.

Giesting, D.W., Roos, M.A. and Easter, R.A. (1991) Evaluation of the effect of fumaric acid and sodium bicarbonate addition on performance of starter pigs fed diets of different types. *Journal of Animal Science* **69**: 2489–2496.

Gildbert, A. and Raa, J. (1977) Properties of a propionic acid/formic acid preserved silage of cod viscera. *Journal of the Science of Food and Agriculture* **28**: 647–653.

Gil de los Santos, J.R., Storch, O.B. and Gil-Turnes, C. (2005) *Bacillus cereus* var. *toyoii* and *Saccharomyces boulardii* increased feed efficiency in broilers infected with *Salmonella enteritidis. British Poultry Science* **46**: 494–497.

Gislason, G., Olsen, R.E. and Ringø, E. (1996) Comparative effects of dietary Na-lactate on Arctic char, *Salvelinus alpinus* L., and Atlantic salmon, *Salmo salar* L. *Aquaculture Research* **27**: 429–435.

Goddeeris, B.M. (2005) Crosstalk between nutrition and immunity. *Proceedings of the Society of Nutrition and Physiology* **14**: 15–20.

Günther, K.D. and Bossow, H. (1998) The effect of etheric oil from *Origanum vulgaris* (Ropadiar®) in the feed ration of weaned pigs on their daily feed intake, daily gains and food utilization. *15th International Pig Veterinarian Society Congress*, Birmingham, United Kingdom, p. 223.

Hardy, R.W. (2000) New developments in aquatic feed ingredients, and potential of enzyme supplements. In: *Avances en Nutrición Acuícola* V. Memorias del V Simposium Internacional de Nutrición Acuícola. 19–22 Noviembre 2000, Mérida, Yucatán, Mexico. Edited by Cruz-Suárez, L.E., Ricque-Marie, D., Tapia-Salazar, M., Olvera-Novoa, M.A., Civera-Cerecedo, R., pp. 216–226.

Hartemink, R. and Rombouts, F.M. (1997) Gas formation from oligosaccharides by the intestinal microflora. In: *Non-Digestible Oligosaccharides: Healthy Food for the Colon? Proceedings of an International Symposium*. Edited by Hartemink, R. pp. 57–66. Graduate School VLAG, Wageningen Institute of Animal Science, Wageningen, The Netherlands.

He, G., Baidoo, S.K., Yang, Q., Gols, D. and Tungland, B. (2002) Evaluation of chicory inulin extracts as feed additiove for early-weaned pigs. Journal of

Animal Science **80** (Suppl. 1): 393.

Helander, I.M., Alakomi, H.L., Latva-Kala, K., Mattila-Sandholm, T., Pol, I., Smid, E.J., Gorris, L.G.M. and von Wright, A. (1998) Characterization of the action of selected essential oil components on Gram-negative bacteria. *Journal of Agricultural and Food Chemistry* **46**: 3590–3595.

Hillman, K., Spencer, R.J., Murdoch, T.A. and Stewart, C.S. (1995) The effect of mixtures of Lactobacillus spp. on the survival of enterotoxigenic *Escherichia coli* in *in vitro* continuous culture of porcine intestinal bacteria. *Letters in Applied Microbiology* **20**: 130–133.

Hopwood, D.E., Pethick, D.W., Pluske, J.R., Hampson, D.J. (2004) Addition of pearl barley to a rice-based diet for newly weaned piglets increases the viscosity of the intestinal contents, reduces starch digestibility and exacerbates post-weaning colibacillosis. *British Journal of Nutrition* **92**: 419–427.

Houdijk, J.G., Bosch, M.W., Tamminga, S., Verstegen, M.W.A., Berenpas, E.B. and Knoop, H. (1999) Apparent ileal and total-tract nutrient digestion by pigs as affected by dietary nondigestible oligosaccharides. *Journal of Animal Science* **77**: 148–158.

Houdijk, J., Bosch, M.W., Verstegen, M.W.A. and Berenpas, H.J. (1998) Effects of dietary oligosaccharides on the growth performance and faecal characteristics of young growing pigs. *Animal Feed Science and Technology* **71**: 35–48.

Hristov, A.N., McAllister ,T.A., van Herk, F.H., Cheng, K.J., Newbold, C.J. and Cheeke, P.R. (1999) Effect of *Yucca schidigera* on ruminal fermentation and nutrient digestion in heifers. *Journal of Animal Science* **77**: 2554–2563.

Huber, J.T. and Soejono, M. (1977) Organic acid treatment of high dry matter corn silage fed to lactating dairy cows. *Journal of Dairy Science* **59**: 2063–2070.

Ikegami, S., Tsuchihashi, F., Harada, H., Tsuchihashi, N., Nishide, E. and Innami, S. (1990) Effect of viscous indigestible polysaccharides on pancreatic-biliary secretion and digestive organs in rats. *Journal of Nutrition* **120**: 353–60.

Inborr, J., Schmitz, M. and Ahrens, F. (1993) Effect of adding fibre and starch degrading enzymes to a barley/wheat based diet on performance and nutrient digestibility in different segments of the small intestine of early weaned pigs. *Animal Feed Science and Technology* **44**: 113–127.

Jensen, B.B. (1988) Effect of diet composition and virginiamycin on microbial activity in the digestive tract of pigs. In: *Proceedings of the 4$^{th}$ International Seminar on Digestive Physiology in the Pig*. Polish Academy of Science, Jablonna, Poland. Edited by Buraczewska, L., Buraczewski, S.,

Pastuszewska, B., Zebrowska, T. pp. 392–400.

Jensen, B.B., Maikkelsen, L.L., Canibe, N. and Høyberg, O. (2001) *Salmonella* in slaughter pigs. Annual Report 2001 from the Danish Institute of Agricultural Science Research Centre Foulum, Ttjele, Denmark, p. 23.

Jeroch, H., Dusel, D., Kluge, H. and Nonn, H. (1999) The effectiveness of microbial xylanase in piglet rations based on wheat, wheat and rye or barley respectively. *Landbauforschung Völkenrode*, FAL, Braunschweig, Germany. Pp. 223–228.

Jin, L.Z., Marquardt, R.R. and Zhao, X. (2000) A strain of *Enterococcus faecium* (18C23) inhibits adhesion of enterotoxigenic *Escherichia coli* K88 to porcine small intestine mucus. *Applied and Environmental Microbiology* **66**: 4200–4204.

Jongbloed, A.W., Kemme, P.A. and Mroz Z. (1996a) Phytase in swine rations: Impact on nutrition and environment. *BASF Technical Symposium*. January 29, 1996, Des Moines, Iowa, BASF, Mount Olive, New Jersey, USA, pp. 44–69.

Jongbloed, A.W., Kemme, P.A., Mroz, Z. and Jongbloed, R. (1996b) The effect of organic acids in diets for growing pigs on the efficacy of microbial phytase. In: *Animal Nutrition and Waste Management*. BASF Reference Manual. Edited by Coelho, M.B., Kornegay, E.T. pp. 515–524.

Juskiewicz, J., Jankowski, J., Zdunczyk, Z., Biedrzycka, E. and Koncicki, A. (2005) Performance and microbial status of turkeys fed diets containing different levels of inulin. *Archiv für Geflügelkunde* **69**: 175–180.

Juven, B.J., Kanner, J., Schved, F. and Weisslovicz, H. (1994) Factors that interact with the antibacterial action of thyme essential oil and its active constituents. *Journal of Applied Bacteriology* **76**: 626–631.

Kamel, C. (2000) Natural plant extracts: Classical remedies bring modern animal production solutions. *3rd Conference on Sow Feed Manufacturing in the Mediterranean Region*. March 22–24, Reus, Spain, pp. 31–38.

Kelly, D. and King, T.P. (2001) Luminal bacteria: regulation of gut function and immunity. In: *Gut environment of pigs*. Edited by Piva, A., Bach Knudsen, K.E., Lindberg, J.E. pp. 113–131. Nottingham University Press, Nottingham, UK.

Kemme, P.A. (1998) Phytate and phytases in pig nutrition: impact on nutrient digestibility and factors affecting phytase efficacy. Doctoral thesis, University of Utrecht, The Netherlands.

Khajarern, J. and Khajarern, S. (2002) The efficacy of origanum essential oils in sow feed. *International Pig Topics* **17**: 17.

Kies, A. and Schutte, J.B. (1997) The effect of microbial phytase on broiler performance. *11th European Symposium on Poultry Nutrition WPSA*. Faaborg, Denmark, pp. 453–455.

Kies, A.K., van Hemert, K.H.F. and Sauer, W.C. (2001) The effect of phytase on protein and amino acid digestibility and energy utilisation. *World's Poultry Science Journal* **57**: 109–126.

Kim, S.W., Knabe, D.A., Hong, K.J. and Easter, R.A. (2003) Use of carbohydrases in corn–soybean meal-based nursery diets. *Journal of Animal Science* **81**: 2496–2504.

Kincaid, R.L., Garikipati, D.K., Nennich, T.D. and Harrison, J.H. (2005) Effect of grain source and exogenous phytase on phosphorus digestibility in dairy cows. *Journal of Dairy Science* **88**: 2893–2902.

Kirchgessner, M. and Roth, F.X. (1988) Ergotrope Effekte durch organische Säuren in der Ferkelaufzucht und Schweinemast. *Übersichten zur Tierernährung* **16**: 93–108.

Kirchegessner, M. and Roth, F.X. (1982) Fumaric acid as a feed additive in pig nutrition. *Pig News and Information* **3**: 259.

Kirchgessner, M., Gedek, B., Wiehler, S., Bott, A., Eidelsburger, U. and Roth, F.X. (1992) Zum Einfluss von Ameisensäure, Calciumformiat und Natriumhydrogencarbonat auf die Keimzellen der Mikroflora und deren Zusammensetzung in verschiedenen Segmenten des Gastrointestinaltraktes. 10. Mitteilung: Untersuchungen zur nutritiven Wirksamkeit von organischen Säuren in der Ferkelaufzucht. *Journal of Animal Physiology and Animal Nutrition* **68**: 73–81.

Klein, U. (1995) Untersuchungen zur Wirksamkeit des Bioregulators Paciflor® in der Kälber- sowie Bullenmast. In: *Vitamine und weitere Zusatzstoffe in der Ernährung von Mensch und Tier.* 5. Symposium, 28./29.09.1995 in Jena/Thüringen. Edited by Schubert, R., Flachowsky, G., Bitsch, R., pp. 494–499.

Knarreborg, A., Jensen, S.K. and Engberg, R.M. (2003) Pancreatic lipase activity as influenced by unconjugated bile acids and pH, measured *in vitro* and *in vivo*. *Journal of Nutritional Biochemistry* **14**: 259–265.

Ko, T.G., Kim, J.D., Bae, S.H., Han, Y.K. and Han, I.K. (2000) Study for the development of antibiotics-free diet for weanling pigs. *Korean Journal of Animal Science and Technology* **42**: 37–44.

Kornegay, E.T. and Qian, H. (1996) Replacement of inorganic phosphorus by microbial phytase for young pigs fed on a maize-soyabean-meal diet. *British Journal of Nutrition* **76**: 563–578.

Kornegay, E.T., Evans, J.L. and Ravindran, V. (1994) Effects of diet acidity and protein concentration or source of calcium on the performance, gastrointestinal content measurements, bone measurements, and carcass composition of gilt and barrow weanling pigs. *Journal of Animal Science* **72**: 2670–2680.

Krause, D.O., Harrison, P.C. and Easter, R.A. (1994) Characterization of the nutritional interactions between organic acids and inorganic bases in the pig and chick. *Journal of Animal Science* **72**: 1257–1262.

Krehbiel, C.R., Rust, S.R., Zhang, G. and Gilliland, S.E. (2003) Bacterial direct-fed microbials in ruminant diets: Performance response and mode of action. *Journal of Animal Science* **81** (E. Suppl. 2): E120–E132.

Kroismayr, A., Sehm, J., Mayer, H., Schreiner, M., Foissy, H., Wetscherek, W. and Windisch, W. (2005) Effect of essential oils or Avilamycin on microbial, histological and molecular–biological parameters of gut health in weaned piglets. *4. BOKU-Symposium Tierernährung: Tierernährung ohne Antibiotische Leistungsförderer.* Vienna, Austria, pp. 140–146.

Kung, L. Jr. (2000) Use of forage additives in silage fermentation. In: *2000-01 Direct-fed Microbial, Enzyme and Forage Additive Compendium*, pp. 39–44. The Miller Publishing Company, Minnesota, USA.

Kung, L. Jr., Treacher, R.J., Nauman, G.A., Smagala, A.M., Endres, K.M. and Cohen, M.A. (2000) The Effect of treating forages with fibrolytic enzymes on its nutritive value and lactation performance of dairy cows. *Journal of Dairy Science* **83**: 115–122

Kung L., Jr., Sheperd, A.C., Smagala, A.M., Endres, K.M., Bessett, C.A., Ranjit, N.K. and Glancey, J.L. (1998) The effect of preservatives based on propionic acid on the fermentation and aerobic stability of corn silage and a total mixed ration. *Journal of Dairy Science* **81**: 1322–1330.

Kyriakis, S.C., Tsiloyiannis, V.K., Vlemmas, J., Sarris, K., Tsinas, A.C., Alexopoulos, C. and Jansegers, L. (1999) The effect of probiotic LSP 122 on the control of post-weaning diarrhoea syndrome of piglets. *Research in Veterinary Science* **67**: 223–228.

Lall, S.P. (1991) Digestibility, metabolism and excretion of dietary phosphorus in fish. In: *Nutritional Strategies and Aquaculture Waste*. Edited by Cowey, C.B., Cho, C.Y. pp. 21–36. Guelph, Ontario, Canada.

Lan, Y., Verstegen, M.W.A., Tamminga, S. and Williams, B.A. (2005) The role of the commensal gut microbial community in broiler chickens. *World's Poultry Science Journal* **61**: 95–104.

Lantzsch, H.-J. (1990) Untersuchungen über ernährungsphysiologische Effekte des Phytats bei Monogastriern (Ratte, Schwein). *Übersichten zur Tierernährung* **18**: 197–212.

Le Blay, G., Blottiere, H.M., Ferrier, L., Le Foll, E., Bonnet, C., Galmiche, J.P. and Cherbut, C. (2000) Short-chain fatty acids induce cytoskeletal and extracellular protein modifications associated with modulation of proliferation on primary culture of rat intestinal smooth muscle cells. *Digestive Diseases and Science* **45**: 1623–1630.

Lee, H.-S. and Ahn, Y.-J. (1998) Growth-inhibiting effects of *Cinnamomum cassia* bark-derived materials on human intestinal bacteria. Journal of

Agricultural and Food Chemistry **46**: 8–12.

Lee, Y.-K., Nomoto, K., Salminen, S. and Gorbach, S.L. (1999) Handbook of Probiotics. John Wiley and Sons, New York.

LeMieux, F.M., Southern, L.L. and Bidner, T.D. (2003) Effect of mannan oligosaccharides on growth performance of weanling pigs. *Journal of Animal Science* **81**: 2482–2487.

Levy, S.B. (2001) Antibiotic resistance: Consequence of inaction. *Clinical Infectious Diseases* **33** (Suppl. 3): S124–S129.

Li, P. and Gatlin, III D.M. (2005) Evaluation of the prebiotic GroBiotic®-A and brewers yeast as dietary supplements for sub-adult hybrid striped bass (*Morone chrysops*×*M. saxatilis*) challenged in situ with *Mycobacterium marinum*. *Aquaculture* **248**: 197–205.

Liebert, F. and Portz, L. (2005) Nutrient utilization of Nile tilapia *Oreochromis niloticus* fed plant based low phosphorus diets supplemented with graded levels of different sources of microbial phytase. *Aquaculture* **248**: 111–119.

Losa, R. (2000) The use of essential oils in animal nutrition. *3rd Conference on Sow Feed Manufacturing in the Mediterranean Region*. March 22–24, Reus, Spain, pp. 39–44.

Lueck, E. (1980) Antimicrobial Food Additives: Characteristics, Uses, Effects. Springer-Verlag, Berlin, Germany.

Lückstädt, C. (2006) Probiotics and premixes in aquaculture – a solution for antibiotic free feeding in shrimp hatcheries in South East Asia. (submitted).

Lückstädt, C. (2005) Organic aquaculture – sustainable production without antibiotic growth-promoters. http://www.efeedlink.com/ (accessed: 18.05.2005).

Lückstädt, C., Senköylü, N., Akyürek, H. and Ağma, A. (2004) Acidifier – a modern alternative for anti-biotic free feeding in livestock production, with special focus on broiler production. *Veterinarija ir Zootechnika* **27**: 91–93.

Mabahinzireki, G.B., Dabrowski, K., Lee, K.J. El-Saidy, D. and Wissner, E.R. (2001) Growth, feed utilization and body composition of tilapia (*Oreochromis* sp.) fed with cottonseed meal-based diets in a recirculating system. *Aquaculture Nutrition* **7**: 189–200.

Maeng, W.J., Kim, C.W. and Shin, H.T. (1987) Effect of a lactic acid bacteria concentrate (*Streptococcus faecium* Cernelle 68) on growth rate and scouring prevention in dairy calves. *Journal of Dairy Science* **9**: 204–210.

Maga, J.A. (1982) Phytate: Its chemistry, occurrence, food interactions, nutritional significance, and methods of analysis. *Journal of Agricultural and Food Chemistry* **30**: 1–9.

Mahan, D.C., Wiseman, T.D., Weaver, E. and Russell, L. (1999) Effect of supplemental sodium chloride and hydrochloric acid added to initial diets containing sprayed-dried blood plasma and lactose on resulting performance and nitrogen digestibility of 3-week-old weaned pigs. *Journal of Animal Science* **77**: 3016–3021.

Maisonnier-Grenier, S., Liu, K., Balasini, M., Dalibard, P. and Geraert, P.A. (2005) Supplementing drinking water with NSP-enzyme: an alternative solution to post-pelleting application. *15th European Symposium on Poultry Nutrition*, 24–29 September 2005, Balatonfüred, Hungary, pp. 369–371.

Männer, K., Jadamus, A., Vahjen, W. and Simon, O. (2002) Effekte probiotischer Zusätze bei Puten auf Leistungsmerkmale und intestinale Mikroflora. *7. Tagung Schweine- und Geflügelernährung*, Lutherstadt Wittenberg, Germany, pp. 78–80.

Manzanilla, E.G., Perez, J.F., Martin, M., Kamel, C., Baucells, F. and Gasa, J. (2004) Effect of plant extracts and formic acid on the intestinal equilibrium of early-weaned pigs. *Journal of Animal Science* **82**: 3210–3218.

Maribo, H., Olsen, L.E., Jensen, B.B. and Miquel, N. (2000) Combination of lactic acid and formic acid and benzoic acid to piglets. Publication no. 490. The National Committee for Pig Production, Copenhagen, Denmark.

Massam, J. (2005) Direct fed microbials in growout ponds: Industrial scale results. *WAS Conference*, 9-13 May 2005, Bali, Indonesia. Abstract CD-Rom.

Mathe, A. (1996) Essential oils as phytogenic feed additives. In: *27th International Symposium on Essential Oils: Essential oils Basic and Applied Research.* Edited by Franz, Ch., Mathe, A., Buchbauer, G. pp. 315–321. Allured Publishing Corporation, Vienna, Austria.

Mathew, A.G., Robbins, C.M., Chattin, S.E. and Quigley, J.D., 3rd (1997) Influence of galactosyl lactose on energy and protein digestibility, enteric microflora, and performance of weanling pigs. *Journal of Animal Science* **75**: 1009–1016.

Matsui, T. (2002) Relationship between mineral availabilities and dietary phytate in animals. *Animal Science Journal* **73**: 21–28.

Mc Cracken, B.A., Gaskins, H.R., Kaiser, R., Klasing, K.C. and Jewell, D.E. (1995) Diet-dependent and diet-independent metabolic responses underlie growth stasis of pigs at weaning. *Journal of Nutrition* **125**: 2838–2845.

Meng, X., Slominski, B.A., Nyachoti C.M., Campbell L.D. and Guenter W. (2005) Degradation of cell wall polysaccharides by combinations of carbohydrase enzymes and their effect on nutrient utilization and broiler chicken performance. *Poultry Science* **84**: 37–47.

Meng, X., Slominski, B.A. and Guenter, W. (2004) The effect of fat type, carbohydrase, and lipase addition on growth performance and nutrient utilization of young broilers fed wheat-based diets. *Poultry Science* **83**:

1718–1727.

Miller, J.A., Solis, L.A. and Laurenz J.C. (2003) Enhancing feed intake during early lactation period in sows. *Journal of Animal Science* **81** (Suppl. 2): 14.

Mitsch, P., Zitterl-Eglseer, K., Köhler, B., Gabler, C., Losa, R. and Zimpernik, I. (2004) The effect of two different blends of essential oil components on the proliferation of *Clostridium perfringens* in the intestines of broiler chickens. *Poultry Science* **83**: 669–675.

Mohnl, M., Hornikova, E., Nitsch, S. and Schatzmayr, G. (2006) Effect of a combination of probiotics, prebiotics and immune-modulating substances on the performance of broiler chickens. *XII European Poultry Conference*, Verona, Italy, 10–14 September 2006 (submitted).

Molnar, O., Schatzmayr, G., Fuchs, E. and Prillinger, H. (2004) *Trichosporon mycotoxinivorans* sp. nov., a new yeast species useful in biological detoxification of various mycotoxins. *Systematic and Applied Microbiology* **27**: 661–671.

Montagne, L., Pluske, J.R. and Hampson, D.J. (2003) A review of interactions between dietary fibre and the intestinal mucosa, and their consequences on digestive health in young non-ruminant animals. *Animal Feed Science and Technology* **108**: 95–117.

Mosenthin, R. and Bauer, E. (2000) The potential use of prebiotics in pig nutrition. *Asian-Australasian Journal of Animal Science* **13**: 315–325.

Mountzouris, K.C., Beneas, H., Tsirtsikos, P., Kalamara, E. and Fegeros, K. (2006) Efficacy of a new multi-strain probiotic product in promoting broiler performance and modulating the composition and activities of cecal microflora. *2006 International Poultry Scientific Forum*, Atlanta, Georgia, p. 59.

Mroz, Z. (2003) Organic Acids of various origin and physico-chemical forms as potential alternatives to antibiotic growth promoters for pigs. *9th International Symposium on Digestive Physiology in Pigs*: Vol. 1. Banff, Alberta, Canada. pp. 267–293.

Mroz, Z., Reese, D.E., Øverland, M., van Diepen, J.T.M. and Kogut, J. (2002) The effects of potassium diformate and its molecular constituents on the apparent ileal and fecal digestibility and retention of nutrients in growing-finishing pigs. *Journal of Animal Science* **80**: 681–690.

Muramatsu, T., Kodama, H., Morishita, T., Furuse, M. and Okumura, J. (1991) Effect of intestinal microflora on digestible energy and fiber digestion in chickens fed a high-fiber diet. *American Journal of Veterinary Research* **52**: 1178–1181.

Namkung, H., Li, M., Gong, J., Yu, H., Cottrill, M. and de Lange, C.F.M. (2004) Impact of feeding blends of organic acids and herbal extracts on growth performance, gut microbiota and digestive function in newly weaned pigs. *Canadian Journal of Animal Science* **84**: 697–704.

New, M. and Csavas, I. (1995) Will there be enough fish meal for fish meals? *Aquaculture Europe* **19**: 6–13.

Nemcova, R., Bomba, A., Gancarcikova, S., Herich, R. and Guba, P. (1999) Study of the effect of *Lactobacillus paracasei* and fructooligosaccharides on the faecal microflora in weanling piglets. *Berliner und Münchener Tierärztliche Wochenschrift* **112**: 225–228.

Ofek, I., Mirelman, D. and Sharon, N. (1977) Adherence of *Escherichia coli* to human mucosal cells mediated by mannose receptors. *Nature* **265**: 623–625.

Ogunkoya, A.E., Page, G.I., Adewolu, M.A. and Bureau, D.P. (2005) Dietary incorporation of soybean meal and exogenous enzyme cocktail can affect physical characteristics of faecal material egested by rainbow trout (*Oncorhynchus mykiss*). *Aquaculture* (in press).

Omogbenigun, F.O., Nyachoti, C.M. and Slominski, B.A. (2004) Dietary supplementation with multienzyme preparations improves nutrient utilization and growth performance in weaned pigs. *Journal of Animal Science* **82**: 1053–1061.

O'Quinn, P.R., Funderburke, D.W. and Tibbetts, G.W. (2001) Effects of dietary supplementation with mannan oligosaccharides on sow and litter performance in a commercial production system. *Journal of Animal Science* **79** (Suppl. 1): 212.

Orban, J.I., Patterson, J.A., Adeola, O., Sutton, A.L. and Richards, G.N. (1997) Growth performance and intestinal microbial populations of growing pigs fed diets containing sucrose thermal oligosaccharide caramel. *Journal of Animal Science* **75**: 170–175.

Overland, M., Granli, T., Kjos, N.P., Fjetland, O., Steien, S.H. and Stokstad, M. (2000) Effect of dietary formates on growth performance, carcass traits, sensory quality, intestinal microflora, and stomach alterations in growing-finishing pigs. *Journal of Animal Science* **78**: 1875–1884.

Oyofo, B.A., DeLoach, J.R., Corrier, D.E., Norman, J.O., Ziprin, R.I. and Mollenhauer, H.H. (1989) Prevention of *Salmonella typhimurium* colonization of broilers with D-mannose. *Poultry Science* **68**: 1357–1360.

Pabst, R., Geist, M., Rothkötter, H.-J. and Fritz, F.J. (1988) Postnatal development and lymphocyte production of jejunal and ileal Peyer's patches in normal and gnotobiotic pigs. *Immunology* **64**: 539–544.

Panigrahi, A., Kiron, V., Puangkaew, J., Kobayashi, T., Satoh, S. and Sugita, H. (2005) The viability of probiotic bacteria as a factor influencing the immune response in rainbow trout *Oncorhynchus mykiss*. *Aquaculture* **243**: 241–254.

Partanen, K.H. (2001) Organic acids – their efficacy and modes of action in pigs. In: *Gut Environment of Pigs.* Edited by Piva, A., Bach Knudsen, K.E., Lindberg, J.E. pp. 201–218. Nottingham University Press, Nottingham, UK.

Partridge, G.G. (2001) The role and efficacy of carbohydrase enzymes in pig nutrition. In: *Enzymes in Farm Animal Nutrition.* Edited by Bedford, M.R., Partridge, G.G. pp. 161–198. CABI Publishing, Wallingford, UK.

Pascual, M., Hugas, M., Badiola, R.I., Monfort, J.M. and Garriga, M. (1999) *Lactobacillus salivarius* CTC2197 prevents *Salmonella enteritidis* colonization in chickens. *Applied and Environmental Microbiology* **65**: 4981–4986.

Patten, J.D. and Waldroup, P.W. (1988) Use of organic acids in broiler diets. *Poultry Science* **67**: 1178–1182.

Perdigón, G., Fuller, R. and Raya, R. (2001) Lactic acid bacteria and their effect on the immune system. *Current Issues in Intestinal Microbiology* **2**: 27–42.

Pettey, L.A., Carter, S.D., Senne, B.W. and Shriver, J.A. (2002) Effects of *â*-mannanase addition to corn-soybean meal diets on growth performance, carcass traits, and nutrient digestibility of weanling and growing-finishing pigs. *Journal of Animal Science* **80**: 1012–1019.

Pfirter, H.P. (2003) AML-Verbot: Alternative Futterzusatzstoffe zur Leistungsverbesserung. In: *Gesunde Nutztiere – Heutiger Stellenwert der Futterzusatzstoffe in der Tierernährung.* Schriftenreihe aus dem Institut für Nutztierwissenschaften., ETH Zürich 24. Edited by Kreuzer, M., Wenk, C., Lanzini, T. pp. 63–71.

Piva, A., Casadei, G. and Biagi G. (2002) An organic acid blend can modulate swine intestinal fermentation and reduce microbial proteolysis. *Canadian Journal of Animal Science* **82**: 527–532.

Pluske, J.R., Siba, P.M., Pethick, D.W., Durmic, A., Mullan, B.P. and Hampson, D.J. (1996) The incidence of swine dysentery in pigs can be reduced by feeding diets that limit the amount of fermentable substrate entering the large intestine. *Journal of Nutrition* **126**: 2920–2933.

Pointillart, A., Fontaine, N. and Thomasset, M. (1984) Phytate phosphorus utilization and intestinal phosphatases in pigs fed low phosphorus: wheat or corn diets. *Nutrition Reports International* **29**: 473–483.

Pongmaneerat, J. and Watanabe, T. (1992a) Utilisation of soybean meal as protein source in diets for rainbow trout. *Nippon Suisan Gakkaishi* **58**: 1761–1773.

Pongmaneerat, J. and Watanabe, T. (1992b) Utilization of soybean meal as protein source in diets for rainbow trout. *Nippon Suisan Gakkaishi* **58**: 1983–1990.

Probert, L. (2004) Role of enzymes in increasing protein digestibility in soya and other oil seeds. http://www.afma.co.za/ (accessed: 14 December 2005).

Quigley, J.D. III, Kost, C.J. and Wolfe, T.A. (2002) Effects of spray-dried animal plasma in milk replacers or additives containing serum and oligosaccharides on growth and health of calves. *Journal of Dairy Science* **85**: 413–421.

Quigley, J.D. III, Drewry, J.J., Murray, L.M. and Ivey, S.J. (1997) Body weight gain, feed efficiency and fecal scores of dairy calves in response to galactosyl-lactose or antibiotics in milk replacers. *Journal of Dairy Science* **80**: 1751–1754.

Rackis, J.J. (1981) Flatulence caused by soya and its control through processing. *Journal of the American Oil Chemists Society* **58**: 503–511.

Radcliffe, J.S., Zhang, Z. and Kornegay, E.T. (1998) The effects of microbial phytase, citric acid, and their interaction in a corn-soybean meal-based diet for weanling pigs. *Journal of Animal Science* **76**: 1880–1886.

Ramakrishna, R.R., Platel, K. and Srinivasan, K. (2003) *In vitro* influence of species and spice-active principles on digestive enzymes of rat pancreas and small intestine. *Food* **47**: 408–412.

Ramli, N., Heindl, U. and Sunanto, S. (2005) Effect of potassium-diformate on growth performance of tilapia challenged with *Vibrio anguillarum*. *WAS Conference*, May 9–13, 2005, Bali, Indonesia. Abstract CD-Rom.

Ranjit, N.K. and Kung, L. Jr. (2000) The effect of *Lactobacillus buchneri*, *Lactobacillus plantarum*, or a chemical preservative on the fermentation and aerobic stability of corn silage. J. Dairy **83**: 526–535.

Recht, J. (2005) Einfluss Seltener Erden in Verbindung mit phytogenen Zusatzstoffen auf Leistungsparameter beim Ferkel. Doctoral thesis, Ludwig-Maximilians-University, Munich, Germany.

Rice, J.P., Radcliffe, J.S. and Pleasant, R.S. (2002) The effect of citric acid alone or in combination with microbial phytase on gastric pH, and P and DM ileal and fecal digestibilities. *Journal of Animal Science* **80** (Suppl. 1): 38.

Richards, J.D., Gong, J. and de Lange, C.F.M. (2005) The gastrointestinal microbiota and its role in monogastric nutrition and health with an emphasis on pigs: Current understanding, possible modulations, and new technologies for ecological studies. *Canadian Journal of Animal Science* **85**: 421–435.

Ringø, E. (1991) Effects of dietary lactate and propionate on growth, and digesta in Arctic charr, *Salvelinus alpinus* (L.). *Aquaculture* **96**: 321–333.

Ringø, E., Olsen, R.E. and Castell, J.D. (1994) Effect of dietary lactate on growth and chemical composition of Arctic charr *Salvelinus alpinus*. *The Journal of the World Aquaculture Society* **25**: 483–486.

Risley, C.R., Kornegay, E.T., Lindemann, M.D., Wood, C.M. and Eigel, W.N.

(1992) Effect of feeding organic acids on selected intestinal content measurements at varying times postweaning in pigs. *Journal of Animal Science* **70**: 196–206.

Ritz, C.W., Hulet, R.M., Self, B.B. and Denbow, D.M. (1995) Growth and intestinal morphology of male turkeys as influenced by dietary supplementation of amylase and xylanase. *Poultry Science* **74**: 1329–1334.

Roberfroid, M.B. (1998) Prebiotics and synbiotics: concepts and nutritional properties. *British Journal of Nutrition* **80** (Suppl. 2): S197–S202.

Rodriguez, E., Han, Y. and Lei, X.G. (1999) Cloning, sequencing, and expression of an *Escherichia coli* acid phosphatase/phytase gene (appA2) isolated from pig colon. *Biochemical and Biophysical Research Communications* **257**: 117–123.

Roth, F.X. and Kirchgessner, M. (1995) Zum Einsatz von Ameisensäure in der Tierernährung. *5. Forum Tierernährung*, BASF AG, Ludwigshafen, Germany, pp. 5–20.

Roth, F.X., Eckel, B., Kirchgessner, M. and Eidelsburger, U. (1992a) Influence of formic acid on pH-value, dry matter content, concentration of volatile fatty acids and lactic acid in the gastrointestinal tract. 3. Communication: Investigations about the nutritive efficacy of organic acids in the rearing of piglets. *Journal of Animal Physiology and Animal Nutrition* **67**: 148–156.

Roth, F.X., Kirchgessner M., Eidelsburger U. and Gedek B. (1992b) Zur nutritiven Wirksamkeit von *Bacillus cereus* als Probiotikum in der Kälbermast. *Agribiological Research* **45**: 294–302.

Russell, T.J., Kerley, M.S. and Allee, G.L. (1996) Effect of fructooligosaccharides on growth performance of the weaned pig. *Journal of Animal Science* **74** (Suppl. 1): 61.

Sakai, M. (1999) Current research status of fish immunostimulants. *Aquaculture* **172**: 63–92.

Samegai, Y. (2000) Enhancement of immune response using a peptidoglycan mixture in pig. *16$^{th}$ International Pig Veterinarian Society Congress*, Melbourne, Australia, p. 176.

Sarica, S., Ciftci, A., Demir, E., Kilinc, K. and Yildirim, Y. (2005) Use of an antibiotic growth promoter and two herbal natural feed additives with and without exogenous enzymes in wheat based broiler diets. *South African Journal of Animal Science* **35**: 61–72.

Sasaki, T., Maede, Y. and Namioka, S. (1987) Immunopotentiation of the mucosa of the small intestine of weaning piglets by peptidoglycan. *Japanese Journal of Veterinary Science* **49**: 235–243.

Schneemann, B.O., Richter, B.D. and Jacobs, L.R. (1982) Response to dietary wheat bran in the exocrine pancreas and intestine of rats. *Journal of Nutrition* **112**: 283–286.

Schoenherr, W.D. (1994) Phosphoric acid-based acidifiers explored for starter diets. *Feedstuffs* **66**: 14.

Sebastian, S., Phillip, L.E., Fellner, V. and Idziak, E.S. (1996) Comparative assessment of bacterial inoculation and propionic acid treatment of aerobic stability and microbial populations of ensiled high-moisture ear corn. *Journal of Animal Science* **74**: 447–456.

Shanklin, R.K. (2001) Effect of form and amount of phosphorus and phytase supplementation on phosphorus utilization by ruminants. MSc thesis, Virginia Polytechnic Institute and State University, Blacksburg, VA, USA.

Shim, S.B. (2005) Effects of prebiotics, probiotics and synbiotics in the diet of young pigs. Doctoral thesis, Wageningen University and Research Centre, Wageningen, The Netherlands.

Shimkus, A. (2004) Prebiotic and synbiotic preparations in calf feeding. *Bulgarian Journal of Agricultural Science* **10**: 491–198.

Shu, Q., Qu, F. and Gill, H.S. (2001) Probiotic treatment using *Bifidobacterium lactis* HN019 reduces weanling diarrhea associated with rotavirus and *Escherichia coli* infection in a piglet model. *Journal of Pediatric Gastroenterology and Nutrition* **33**: 171–177.

Simon, O. (2005) Mikroorganismen als Futterzusatzstoffe: Probiotika – Wirksamkeit und Wirkungsweise. *4. BOKU-Symposium Tierernährung: Tierernährung ohne Antibiotische Leistungsförderer.* Vienna, Austria, pp. 10–16.

Simon, O. (1998) The mode of action of NSP hydrolysing enzymes in the gastrointestinal tract. *Journal of Animal and Feed Sciences* **7**: 115–123.

Sinlae, M. and Choct, M. (2000) Xylanase supplementation affects the gut microflora of broilers. *Australian Poultry Science Symposium.* Sydney, Australia, pp. 209.

Sissons, J.W. (1989) Potential of probiotic organisms to prevent diarrhoea and promote digestion in farm animals. A review. *Journal of the Science of Food and Agriculture* **49**: 1–13.

Skinner, J.T., Izat, A.L. and Waldroup, P.W. (1991) Research note: Fumaric acid enhances performance of broiler chickens. *Poultry Science* **70**: 1444–1447.

Steiner, T., Kroismayr, A. and Zhang, C. (2006a) Evaluation of a phytogenic feed additive as natural growth promoter for piglets. *21st International Pig Veterinarian Society Congress*, Copenhagen, Denmark (in press).

Steiner, T., Mosenthin, R., Fundis, A. and Jakob, S. (2006) Influence of feeding level on apparent total tract digestibility of phosphorus and calcium in pigs fed low-phosphorus diets supplemented with microbial or wheat phytase. *Livestock Science* **102**: 1-10.

Steiner, T., Mosenthin, R., Zimmermann, B., Greiner, R. and Roth, S. (2006c) Distribution of phytase activity, total phosphorus and phytate phosphorus in legume seeds, cereals and cereal by-products as influenced by harvest year and cultivar. *Animal Feed Science and Technology* (in press).

Stokes, C.R., Bailey, M. and Haverson, K. (2001) Development and function of the pig gastrointestinal immune system. In: *Proceedings of the 8th Symposium on Digestive Physiology of Pigs*. Edited by Lindberg, J.E., Ogle, B. pp. 59–66. CAB International, Wallingford, UK.

Sugiura, S.H., Gabaudan, J., Dong, F.M. and Hardy, R.W. (2001) Dietary microbial phytase supplementation and the utilization of phosphorus, trace minerals and protein by rainbow trout *Oncorhynchus mykiss* (Walbaum) fed soybean meal-based diets. *Aquaculture* Research **32**: 583–592.

Swann, M.M. (1969) Use of Antibiotics in Animal Husbandry and Veterinary Medicine. UK Joint Committee Report London: H.M. Stationery Office.

Taylor, D.J. (1999) The responsible use of antibiotics in pig medicine. *The Pig Journal* **43**: 170–187.

Teas, J., Harbison, M.L. and Gelman, R.S. (1984) Dietary seaweed (*Laminaria*) and mammary carcinogenesis in rats. *Cancer Research* **44**: 2758–2761.

Thaela, M.J., Jensen, M.S., Pierzynowski, S.G., Jakob, S. and Jensen, B.B. (1998) Effect of lactic acid supplementation on pancreatic secretion in pigs after weaning. *Journal of Animal and Feed Sciences* **7** (Suppl. 1): 181–183.

Turner, J.L., Dritz, S.S., Higgins, J.J. and Minton, J.E. (2002) Effects of *Ascophyllum nodosum* extract on growth performance and immune function of young pigs challenged with *Salmonella typhimurium. Journal of Animal Science* **80**: 1947–1953

Underdahl, N.R. (1983) The effect of feeding *Streptococcus faecium* upon *Escherichia coli* induced diarrhea in gnotobiotic pigs. *Progress in Food and Nutrition Science* **7**: 5–12.

Vahjen, W., Gläser, K., Schäfer, K. and Simon, O. (1998) Influence of xylanase-supplemented feed on the development of selected bacterial groups in the intestinal tract of broiler chicks. *Journal of Agricultural Science* **130**: 489–500.

Van Heugten, E. and van Kempen, T. (2001) Understanding and applying nutrition concepts to reduce nutrient excretion in swine. College of Agriculture and Life Sciences, North Carolina State University. http://mark.asci.ncsu.edu/Nutrition/Environ/concepts.pdf. (accessed: 22.09.2003).

Velàzquez, G., Borbolla, A.G., Reis De Souza, T. and Mariscal, G. (2004) Productive performance of weaned pigs fed diets containing oregano, cinnamon and pepper extracts as growth promoters. *18th International*

*Pig Veterinarian Society Congress*, Hamburg, Germany, p. 767.

Verbeeke, W. (2001) The influence of consumerism on livestock production and eventually the feed industry. *Animal Feed Manufactures Association (AFMA) Forum 2001.* Recent Developments in animal feeds and feeding. Sun City, Northwest Province, South Africa, pp. 1–18.

Viveros, A., Brenes, A., Pizzaro, M. and Castano, M. (1994) Effect of enzyme supplementation of a diet based on barley, an autoclave treatment, on apparent digestibility, growth performance and gut morphology of broilers. *Animal Feed Science and Technology* **48**: 237–251.

Vogt, H., Matthes, S. and Harnisch, S. (1981) Der Einfluß organischer Säuren auf die Leistungen von Broilern und Legehennen. *Archiv für Geflügelkunde* **45**: 221–232.

Waldenstedt, L. (2003) Effect of vaccination against coccidiosis in combination with an antibacterial oregano (*Origanum vulgare*) compound in organic broiler production. *Acta Agriculturae Scandinavica Section A-Animal Science* **53**: 101–109.

Walsh, M.C., Peddireddi, L. and Radcliffe, J.S. (2004) Acidification of Nursery Diets and the role of diet buffering capacity. *2004 Swine Research Report.* Purdue University. pp. 25–36.

Walsh, M., Sholly, D., Kelly, D., Cobb, M., Trapp, S., Hinson, R., Hill, B. Sutton, A., Radcliffe, S., Harmon, B., Smith, J. and Richert, B. (2003) The effects of supplementing weanling pigs diets with organic and inorganic acids on growth performance and microbial shedding. *2003 Swine Research Report*. Purdue University, pp. 89–98.

Walton, J.R. (1983) Modes of action of growth promoting agents. *Veterinary Research Communications* **7**: 1–7.

Wang, W.-S. and Wang, D.-H. (1997) Enhancement of the resistance of tilapia and grass carp to experimental *Aeromonas hydrophila* and *Edwardsiella tarda* infections by several polysaccharides. *Comparative Immunology, Microbiology and Infectious Diseases* **20**: 261–270.

Webb, P.R., Kellogg, D.W., Mcgahee, M.W. and Johnson, Z.B. (1992) Addition of fructooligosaccharide (FOS) and sodium diacetate (SD) plus decoquinate (D) to milk replacer and starter grain fed to Holstein calves. *Journal of Dairy Science* **75** (Suppl. 1): 300.

Weinberg, Z.G., Muck, R.E. and Weimer, P.J. (2003) The survival of silage inoculant lactic acid bacteria in rumen fluid. *Journal of Applied Microbiology* **94**: 1066–1071.

White, L.A., Newman, M.C., Cromwell, G.L. and Lindemann, M.D. (2002) Brewers dried yeast as a source of mannan oligosaccharides for weanling pigs. *Journal of Animal Science* **80**: 2619–2628.

Williams, P. and Losa, R. (2001) The use of essential oils and their compounds in poultry nutrition. *World Poultry* **17**: 14–15.

Xu, Z.R., Hu, C.H., Xia, M.S., Zhan, X.A. and Wang, M.Q. (2003) Effects of dietary fructooligosaccharide on digestive enzyme activities, intestinal microflora and morphology of male broilers. *Poultry Science* **82**: 1030–1036.

Yang, W.Z., Beauchemin, K.A. and Rode, L.M. (1999) Effects of an enzyme feed additive on extent of digestion and milk production of lactating dairy cows. *Journal of Dairy Science* **82**: 391–403.

Yamamoto, I., Nagumo, T., Fujihara, M., Takahashi, M. and Ando, Y. (1977) Antitumor effect of seaweeds. II. Fractionation and partial characterization of the polysaccharide with antitumor activity from *Sargassum fulvellum. Japanese Journal of Experimental Medicine* **47**: 133–140.

Yasar, S. and Forbes, J.M. (2000) Enzyme supplementation of dry and wet wheat-based feeds for broiler chickens: performance and gut responses. *British Journal of Nutrition* **84**: 297–307.

Yin, Y.-L., Baidoo, S.K., Schulze, H. and Simmins, P.H. (2001a) Effects of supplementing diets containing hulless barley varieties having different levels of non-starch polysaccharides with â-glucanase and xylanase on the physiological status of the gastrointestinal tract and nutrient digestibility of weaned pigs. *Livestock Production Science* **71**: 97–107.

Yin, Y. L., Baidoo, S.K., Jin, L.Z., Liu, Y.G., Schulze, H. and Simmins, P.H. (2001b) The effect of different carbohydrase and protease supplementation on apparent (ileal and overall) digestibility of nutrients of five hulless barley varieties in young pigs. *Livestock Production Science* **71**: 109-120.

Zani, J.L., Weykamp da Cruz, F., Freitas dos Santos, A. and Gil-Turnes, C. (1998) Effect of probiotic CenBiot on the control of diarrhoea and feed efficiency in pigs. *Journal of Applied Microbiology* **84**: 68–71.

Zdunczyk, Z., Jankowski, J. and Juskiewicz, J. (2005) Performance and intestinal parameters of turkeys fed a diet with inulin and oligofructose. *Journal of Animal and Feed Sciences* **14** (Suppl. 1): 511–514.

Zhan, X.A., Hu, C.H. and Xu, Z.R. (2003) Effects of fructooligosaccharide on growth performance and intestinal microflora and morphology of broiler chicks. *Chinese Journal of Veterinary Science* **23**: 196–198.

Zhang, Z., Marquardt, R.R., Guenter, W. and Han, Z. (1997) Effect of different enzyme preparations supplemented in a rye-based diet on the performance of young broilers and the viscosity of digesta and cloacal excreta. *Chinese Journal of Animal Science* **34**: 3–6.

Zimmermann, B., Lantzsch, H.-J., Mosenthin, R., Schöner, F.-J., Biesalski, H.K. and Drochner, W. (2002) Comparative evaluation of the efficacy of cereal and microbial phytases in growing pigs fed diets with marginal phosphorus supply. *Journal of the Science of Food and Agriculture* **82**: 1298–1304.

Zoetendal, E.G., Collier, C.T., Koike, S., Mackie, R.I. and Gaskins, H.R. (2004) Molecular ecological analysis of the gastrointestinal microbiota: A review. *Journal of Nutrition* **134**: 465–472.

Zrustova, J., Ritter, M., Svoboda, K.P. and Brooker, J.D. (2005) Secondary plant metabolites to control growth of *Clostridium perfringens* from chickens. *15th European Symposium on Poultry Nutrition*. 24–29 September 2005, Balatonfüred, Hungary, pp. 221–233.

# 14

# INDEX

## A



## B



## C

## M



## N



## O



## P

## R



## S



## T



## V



## X



## Y